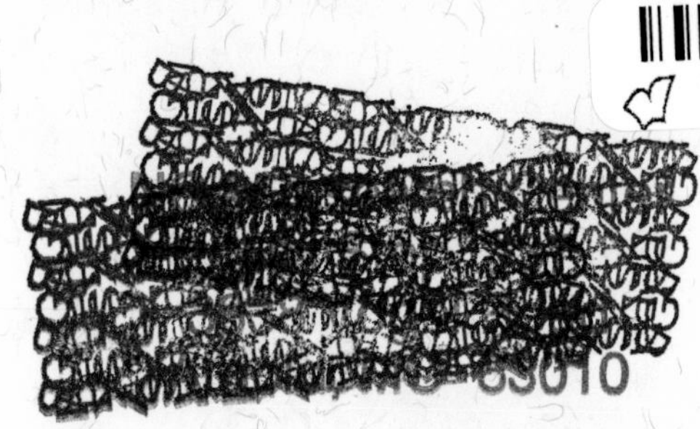

A LONG
GOODBYE

A LONG GOODBYE

The Soviet Withdrawal from Afghanistan

ARTEMY M. KALINOVSKY

HARVARD UNIVERSITY PRESS
Cambridge, Massachusetts, and London, England
2011

Printed in the United States of America

Library of Congress Cataloging-in-Publication Data

Kalinovsky, Artemy M.
A long goodbye : the Soviet withdrawal from Afghanistan / Artemy M. Kalinovsky.
p. cm.
Includes bibliographical references and index.
ISBN 978-0-674-05866-8 (alk. paper)
1. Soviet Union—Foreign relations—Afghanistan. 2. Afghanistan—Foreign relations—Soviet Union. 3. Afghanistan—History—Soviet occupation, 1979–1989. 4. Soviet Union—Foreign relations—1975–1985. 5. Soviet Union—Foreign relations—1985–1991. 6. Gorbachev, Mikhail Sergeevich, 1931– 7. Shevardnadze, Eduard Amvrosievich, 1928– 8. Disengagement (Military science)—Case studies. I. Title.
DK68.7.A6K35 2011
958.104′5—dc22 2010048681

To my parents and grandparents

CONTENTS

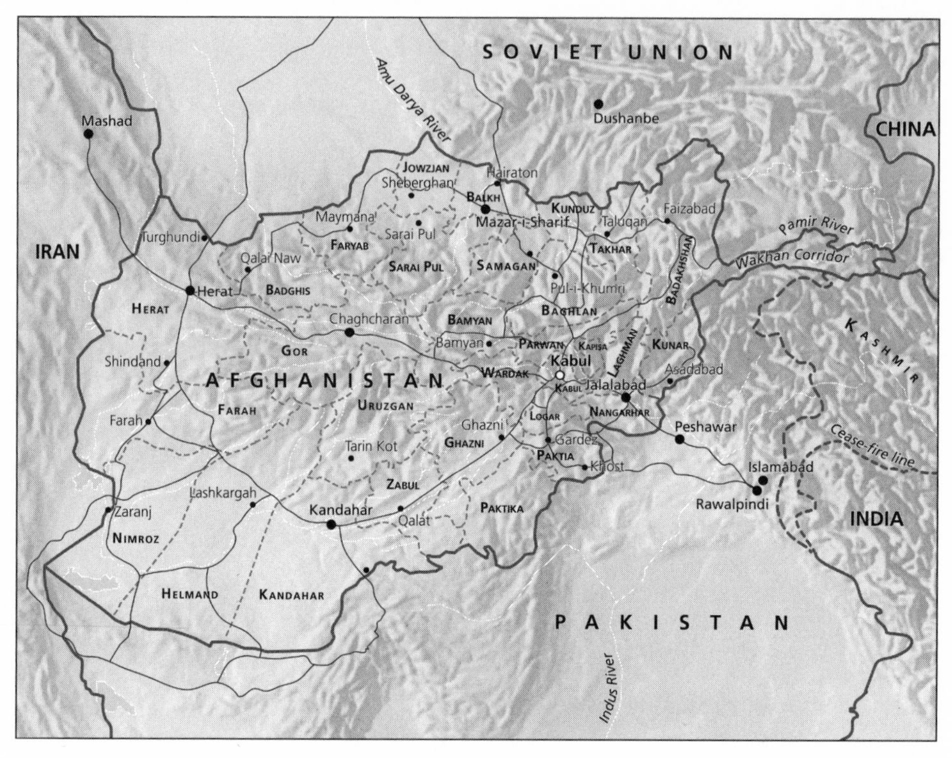

Afghanistan and its neighbors, 1947–1991

INTRODUCTION

The Soviet intervention in Afghanistan (1979–1989) was easily one of the bloodiest of Cold War conflicts. By the time the last Soviet soldier returned to his native soil in February 1989, over 13,000 of his comrades had fallen in the Afghan dirt, and another 40,000 were wounded. Countless more returned home scarred by the war, in ways that sometimes manifested themselves only years later. Estimates of Afghan losses vary, but it is believed that anywhere from 800,000 to 1.2 million Afghans died as a result of the fighting. Yet the carnage did not end there. In the years 1989–1992, the government of the Republic of Afghanistan, supported by Soviet advisers and armaments, continued to hold out against *mujahadeen* groups backed by Pakistan, the United States, and Saudi Arabia. The Republic of Afghanistan outlasted the Soviet Union by almost five months, but its collapse merely precipitated a new phase of civil war, which in one form or another continues to the present day.

The intervention in Afghanistan was the culmination of the Soviet Union's involvement with the Third World that began in the 1950s and extended throughout the 1960s and 1970s. In terms of military aid alone, Soviet advisers during the latter decade took part in the Egyptian-Israeli War of Attrition (1970), the Angolan civil war (from 1975), and the Somali-Ethiopian conflict in the Ogaden Desert (1977–1978).[1] In the European theater of the Cold War, Soviet leaders had also chosen military intervention when communist regimes were threatened in Hungary (1956) and Czechoslovakia (1968). Broadly speaking, all of these involvements were undertaken to shore up Soviet-friendly regimes and demonstrate Moscow's willingness to use force on behalf of allies. Yet while the intervention in Afghanistan had its precedents, it also became a turning point. The war, so costly in blood and treasure, forced Soviet leaders to reevaluate interventions as instruments of foreign policy. Thus, when a crisis broke out in Poland in the summer of 1980, threatening the regime in Warsaw, even an ardent pro-interventionist like KGB chairman Yurii Andropov conceded that "the quota of interventions abroad has been exhausted."[2]

Historians approaching the Afghan war in the years since 1992 have focused primarily on understanding why the Soviet Union intervened in the first place, or on the military aspects of the intervention. This book poses a different question: Why did it take the Soviet Union so long to bring its troops home? After all, shortly after the invasion Soviet leaders realized that the intervention was becoming a quagmire, with serious costs for their relationship with the rest of the world. This question is particularly important because it relates not only to the war in Afghanistan but also to the debate about changes in Soviet foreign-policy thinking in the 1980s, the emergence and influence of the so-called New Political Thinking, and the potential of superpower cooperation and UN involvement in resolving regional conflicts.

This book is therefore first and foremost a study of Soviet decision making. It examines the political struggles behind the troop withdrawal within the Politburo and other institutions involved in the Soviet Union's foreign-policy process, including the military and the KGB. In seeking to understand why certain policies triumphed over others and why key decisions were made (or delayed), this book will analyze the impact of ideology, political legacy, patron-client relations, superpower diplomacy, and bureaucratic politics on elite decision making during the Afghan war.

The single most important reason that Soviet leaders delayed the decision to withdraw for as long as they did is that they continued to believe the USSR could help stabilize Afghanistan, build up the Afghan armed forces, and make the Kabul government more acceptable to its people. This is hardly the only reason, however. Soviet leaders found it difficult to disengage from the Afghan conflict because they feared undermining Moscow's status as a defender of Third World countries against encroaching neo-colonialism. Efforts at withdrawal were complicated by Cold War tensions (particularly in 1980–1985), divisions among Soviet officials and agencies involved in Afghanistan, and the persistence of an "Afghan lobby" within the Soviet leadership that refused to concede defeat in Afghanistan.

Compared to the literature on the Vietnam war, the historiography on the Soviet intervention in Afghanistan has been paltry.[3] There are a number of complex reasons for this, among them the fact that Soviet sources on the war are still difficult to access, that American sources have only recently started to become available, and that research in Afghanistan itself is virtually impossible.[4] A truly international history of the war—one which incorporates the narratives of Washington, Moscow, Kabul, Tehran, Delhi, Riyadh, Beijing, the Afghan resistance, the Arab volunteers, the CIA (America's Central Intelligence Agency), the ISI (Pakistan's Inter-Services Intelligence Agency), the KGB (the USSR's national security agency), and the Afghan refugees, to name a few of the actors in this drama—is at this point very far away. I have set myself a much simpler and less ambitious (although, I believe, still fundamental) task: to understand how Soviet leaders came to accept the need to withdraw, how they went about it, and what obstacles they faced. I chose this angle not only to shed light on a key aspect of the war, but also to engage with the broader debates on intervention, on changes in Soviet strategic thinking, and on the end of the Cold War.

Yet even with this relatively simple topic, it is impossible to tell the story just from Moscow's point of view. After all, as Soviet leaders made their decisions about the Afghan war, they were trying to take into account the position of their allies, the Afghan communists; their standing relative to their rivals, the Americans (and their relations with the same); the opinion of their fellow citizens; and the views of their various allies elsewhere in the Third World.

The Afghan revolution and the subsequent Soviet intervention came at what proved to be a pivotal moment in the history of the twentieth century and that of the Cold War, at least for the actors involved. For the Soviets themselves, the revolution proved the culmination of a decade where Soviet influence had seen some reverses (most notably, Anwar Sadat's aligning of Egypt more closely with the United States) but also some crucial victories in the Third World, including socialist revolutions in Ethiopia and Angola and the US defeat in Vietnam. Moscow had been actively engaged in this struggle since the 1950s. Soon after emerging as the most powerful figure at the top of the Central Committee of the Communist Party of the Soviet Union (CC CPSU), Nikita Khrushchev criticized Stalin's inattention to the Third World in the worldwide communist struggle. With rapid decolonization changing the political map of the world, Moscow became increasingly involved in the transformations taking place in the Third World, including Afghanistan, where it helped King Zahir Shah and his prime minister, Mohammed Daoud, modernize the infrastructure.[5] The competition in the Third World became a three-way contest. Moscow had to prove not only that communism was a better path to modernity than liberal capitalism, but that Soviet communism was better than the Chinese model. In the 1970s, the Soviet Union would find itself involved in a number of situations where the US and China were both backing the rivals of a Moscow ally.[6] (Soviet aid continued into the 1980s. From 1982 to 1986, the USSR provided $78 billion in arms to developing states. Some of the biggest recipients were Ethiopia, Angola, Vietnam, North Korea, and Cuba. India, Syria, and Libya were also major recipients.)[7] The Soviet model seemed to be winning adherents in the colonial world, even as the United States was increasingly identified with new imperialism, inequality, and Zionism, and as its attempts to win allies in the Third World were often handicapped by racism and the legacy of slavery at home.

By the mid-1970s, Soviet leaders felt as if they had finally found a modus vivendi with their American and Western European counterparts. Years of détente diplomacy had brought relations to a certain level of normalcy, and the danger of nuclear war seemed to have receded. Moreover, détente offered Soviet leaders the recognition of parity they craved. (It is worth remembering that even among those leaders who ultimately voted for intervention in Afghanistan, there were passionate believers in détente, not least Leonid Brezhnev himself.) Domestically, Soviet citizens enjoyed standards of living that were lower than those of their Western European and American counterparts, but far above and beyond what they would have remembered from the 1930s, 1940s, and early 1950s; there was censorship and enforced conformity, but state-sponsored terror and mass repression were things of the past. There were serious underlying problems with the economy, some of them a result of the overemphasis on the defense industry at the expense of other sectors, but these problems could be papered over with the foreign reserves the country earned from its export of oil and gas. There was widespread cynicism, but also pride in Soviet achievements, the victory over fascism in the Great Patriotic War, and the country's growing influence around the world.[8]

For the United States, the 1970s were a much more tumultuous decade. The Vietnam war had opened fissures in American society which still had not closed. The foreign-policy consensus was increasingly under attack from both left and right. With rising inflation in the wake of the 1973 oil embargo, widespread unemployment, and high rates of urban crime, the domestic "new deal" political alignment seemed under threat as well. Détente (as practiced by President Richard Nixon and Secretary of State Henry Kissinger) had in some ways been an easier sell abroad than at home, where it was subject to criticism for ignoring human rights and for allowing the Soviet Union to increase its power and influence even as that of the United States waned. The seemingly arrogant Soviet march across the Third World throughout the decade only fed the criticism of both conservative hawks like Ronald Reagan and liberal Cold Warriors like Senator Henry "Scoop" Jackson, Democrat of Washington State.[9]

Both criticisms of détente were shared by President Jimmy Carter, who came to office in 1977, and his National Security Adviser, Zbigniew Brzezinski. Not only would the United States now take more of an interest in human rights behind the Iron Curtain (to the chagrin of Soviet leaders), but Washington would no longer tolerate Soviet advances in the Third World, seen by Brzezinski and many others as a violation of the spirit of détente, a threat to US interests, and ultimately a source of instability. Support for socialist Somalia against Soviet-aligned Ethiopia during the Ogaden war of 1977 thus foreshadowed the Carter administration's response to the Soviet intervention and its support of the *mujahadeen*. The ghost of Vietnam hung over all of these issues. For the much more hawkish Reagan administration, which took office in 1981, and for many of the Cold Warriors in Congress who supported US aid to the *mujahadeen*, the primary motivation may have been to teach Moscow a lesson about acknowledging limits. But the desire to deal communism a blow at the hands of "freedom fighters" and to seek revenge for the defeat in Vietnam was what really gave US involvement in Afghanistan the feel of a crusade, although one that was fought at some distance. The Soviet Union's difficulties in pacifying the Afghan opposition, the strain the intervention put on Moscow's relationship with Third World states, and the possibility of exorcising the ghost of Vietnam all served as a balm for American politicians and Cold Warriors of various stripes who feared, as the 1970s came to a close, that their country and their way of life were in decline.[10]

The intervention was the culmination of a decade marked by major political shocks and realignments in the region. Indian democracy suffered its greatest crisis since the country's independence in 1975, when strikes and political unrest led Indira Gandhi to push for a State of Emergency which lasted until 1977. In Islamabad, the Pakistan People's Party (PPP) had come to power following the country's defeat in 1971 and its loss of West Pakistan (Bangladesh), ending a period of military rule and ushering in a number of progressive reforms. But by 1977 Zulfikar Ali Bhutto had made many enemies—on the right and among business owners for his nationalization of various enterprises, and on the left for cozying up to Pakistan's landholders. Although the PPP did very well in the March 1977 elections, the opposition soon took to the streets. In July, Bhutto and his government were ousted in a coup led by army general Zia ul-Haq. Zia, a devout Muslim who moved the country away from Bhutto's secularizing policies and economic program.[11] In Afghanistan itself, the reformist King Zahir Shah was overthrown by his cousin Mohammed Daoud in 1973. Daoud, a modernizer who brought his country closer to the USSR, ruled with the support of leftists, including the more moderate communists. He sought aid from Iran, Saudi Arabia, Iraq, and Kuwait.[12] To Pakistan's chagrin, he also revived the idea that all the Pashtun lands, which straddled the border between Afghanistan and Pakistan, should be united.[13]

Zia's coup precipitated yet another crisis in US-Pakistani relations. The Carter administration, with its emphasis on human rights, quickly emerged as a critic of Zia. Pakistan's nuclear program (begun by Bhutto to balance India's nuclear capability) was also an increasing irritant, and led the US to cut off aid. Bhutto's execution in 1979 and Pakistan's withdrawal from the Central Treaty Organization brought the alliance to its nadir. After students seized the US Embassy in Islamabad in November 1979 and Pakistani authorities were slow to respond, US relations with Pakistan were, in the words of a National Security Council staff member, "about as bad as with any country in the world, except perhaps Albania or North Korea."[14] What salvaged the relationship in subsequent months was the Soviet intervention in Afghanistan, which brought the US and Pakistan closer together. Zia, who hosted Afghan refugees and emerged as the godfather of the Afghan *jihad,* became a hero instead of a pariah, and for most of the 1980s US politicians were more than happy to overlook his own violations of civil rights, let alone Pakistan's nuclear program.[15] No wonder that when a deal on Soviet withdrawal seemed within reach, Zia worried that this would mean the end of his special relationship with the United States.

The devout Zia's seizure of power was a harbinger of the general turn away from secular-leftist solutions to problems of development in the Muslim world and toward a religiously oriented politics, often referred to today under the catch-all term "political Islam." Disappointment with the modernization programs and attempts at unity by leaders like Nasser in Egypt or the Ba'athists in Syria and Iraq led Arab intellectuals toward solutions based more on Islam than on secular modernity. Arab failures in conflicts against Israel in 1967 and 1973 reinforced the sense that secular Arab nationalism was bankrupt. In Islam these intellectuals found the possibility of unity, resistance to Western imperialism, and provisions for mutual welfare that had eluded them since independence. Similarly, in Iran, young people frustrated with the shah's modernization program (the "White Revolution") and his repressive, corrupt rule turned increasingly to religious thinkers and leaders like Ayatollah Ruhollah Khomeini, who promised a more just and equal society under the law not of a secular despot, but of God. To the Islamist intellectuals as well as their supporters, the intervention in Afghanistan was the latest and perhaps most egregious example of Western imperialist arrogance, a forced secular modernism that was being imposed from outside and that they were duty-bound to fight. The withdrawal of Soviet forces was only an intermediate goal; what they wanted was the defeat of the communist regime in Afghanistan and its replacement by an Islamic state.[16]

Against this backdrop, the Afghan communists who took power in 1978, with their vision of a radical restructuring of Afghan society and massive, rapid modernization, were truly swimming against the current. For many postcolonial leaders, modernization was a lodestar, a path for their nations out of the legacy of dependence and domination they had endured. (In the words of Ghanaian leader Kwame Nkrumah, a once-dependent country had to be "jet propelled" into modernity or "lag behind and thus risk everything for which it has fought.")[17] But the chosen paths of these leaders, whether leaning left or right, rarely had the sustained, wholehearted supported of the population, particularly the peasantry. The construction of the modern state ultimately meant the intrusion of the state into an area where it previously had a "light footprint." In a country like Afghanistan, where land distribution was fairly egalitarian, the revolution's promises rang hollow, and the government representative who came to bring freedom and equality was suspect.[18] The suspicion was confirmed when that representative, accompanied perhaps by a Russian adviser, began to redistribute land, force women and children into schools, and issue antireligious propaganda.

Like their counterparts in other Third World countries, the Afghan revolutionaries came from an intellectual class that was already small and isolated from the rest of society. Many had studied abroad, becoming exposed not only to the achievements and living conditions of the modern world, but also to Marxist ideas and literature. A communist party, the People's Democratic Party of Afghanistan (PDPA), was founded in 1965 by a group of such intellectuals. Nur Muhammad Taraki and Babrak Karmal immediately emerged as its leaders, but disagreements between the two and among their respective supporters led to a split in the party only two years later, in 1967. The two wings that emerged became known as Khalq ("Masses") and Parcham ("Banner"), after their respective newspapers. (Over the years, Soviet representatives tried to broker a reconciliation, but with little success.)[19] What they shared was an embarrassment with their own country's backwardness and a fervent desire to see it catch up in as short a time as possible. A Soviet diplomat who in the 1980s gave a tour of Moscow to Najib, head of the Afghan secret police and later the country's president (as Mohammad Najibullah), recalled his wistful comment: "When will Afghanistan have buildings like these?" To this end, the revolutionaries were willing to go much further than the reformist leaders who, since the 1950s, had gently steered Afghanistan toward more secular ways. They accepted US and Soviet aid for various modernization projects.[20] They also shared a faith in the Soviet Union and its ability to help deliver the kind of change they wanted for their own country. When a young activist, recently returned from a stay in the USSR, began to criticize lack of free speech there and to urge her party leadership to become more independent of Moscow, she was told that "a mosquito on the other side of the river is a fighter jet for us."[21] Small wonder that, when faced with serious uprisings less than a year into their revolution, these leaders began calling on Moscow to send in troops: surely the mighty Soviet Union could handle Afghanistan's problems!

It is perhaps not surprising, then, that the Afghan communists quickly found themselves alone at the top when they took power. The purge of other secular leftists, as well as the Khalqi consolidation of control and repression or forced exile of most Parcham members, was part of the reason. More important, what they attempted to do was fundamentally at odds with the wishes of most sectors of Afghan society, and with the way politics, even modernizing politics, was conducted in Afghanistan. Starting with Amir Abdur Rahman (reigned 1880–1901), Afghan rulers had slowly imposed formal state control on regions of the country, accepting foreign aid while being careful to keep foreign influence to an absolute minimum. (Abdur Rahman's predecessors had tried to do this as well, but he was the first to have significant success in this regard.) They offered opportunities for education to a select few, regularized tax collection, and gradually eroded the autonomy of mullahs and tribal leaders.[22] The revolutionary regime's program, by contrast, called for a radical redistribution of land, banned the burqa, and closed mosques. These were not abstract goals for a future Afghanistan; the Afghan communists, or more specifically the Khalqis, believed in effecting immediate change. Their plans in 1978, if successful, would have completely changed rural society, breaking the authority of traditional leaders and tying all producers to the state. The regime's decrees hit directly at the nexus of long-standing economic and social relations, affecting issues such as bride-price, marriage age, tenancy, and mortgages.[23]

Like modernizing revolutionaries in other traditional societies, the Afghan communists were out of touch with most of their fellow countrymen, but they accepted the imperative to effect the changes they saw as objectively necessary. If there was opposition, it was inspired by reactionary elements, tribal leaders, and the CIA. Repression was thus a justified tool in the service of the revolution. When Soviet leaders complained that the Afghan revolution was moving too fast, the Afghans cited Russia's own history back to them: Hadn't Stalin also pulled their country from the Dark Ages into modernity, refusing to flinch at the human cost and ignoring colleagues who told him to slow down? Although the leaders installed by Moscow at the end of 1979 were more pragmatic and less radical, they were broadly committed to similar ideals. They too would prove reluctant when Soviet advisers tried to cajole them into reversing some aspects of the revolution, returning land, and accepting markets.

The regime's opponents were a varied group. The regime did face spontaneous peasant and even urban resistance, but the main concern of both Kabul and Moscow was the organized militant opposition. The leaders of the Sunni religious opposition in particular shared some traits with the communists: exposure to foreign ideas (in this case, the political Islam of the Egyptian writer Sayyid Qutb, who had been translated into Persian) and politicization in a university setting. This fairly describes Sebghatullah Mujadidi, Barhanuddin Rabbani, Ahmad Shah Massoud, Gulbudin Hekmatyar, and Yunus Khalis. These activists had opposed the more moderate reforms of Mohammed Daoud (president 1973–1978) and even of King Zahir Shah. By the time of the Saur Revolution they were already exiles, forced to escape after a failed uprising in 1975. (Hekmatyar's Hizb-e Islami was the most radical and impatient faction, pushing for a quick attempt at power. Disagreements with Rabbani, Massoud, and the other leaders kept the groups divided and often at cross-purposes throughout the war.) But the intervention and the ensuing refugee crisis provided them with recruits and mass support, as well as with crucial aid from foreign sponsors.[24]

Religious fervor hardly explains the resistance, however. Among the various groups fighting to bring down the Kabul government were royalists and Maoists. Many groups operated with only vague allegiance to any particular leader or ideology. And even among the "Islamist" groups, the motivations of the fighters varied. More than a few were from families that had supported the reforms of Mohammed Daoud; some had even welcomed the Saur Revolution. But the radicalism and the Red Terror those revolutionaries brought threw these people into the arms of the opposition, often after they or their families had been arrested by the regime and had spent time in one of its prisons.[25]

International-relations scholars of the realist school generally agree that interventions are undertaken to prevent or reverse perceived losses.[26] This is certainly true for the Soviet intervention in Afghanistan and elsewhere; but when it comes to describing Cold War interventions, it is also insufficient. The Cold War was more than a geo-strategic confrontation—it was also a competition between two ways of life and two paths to modernity. As Third World leaders embarked on their own experiments and pledged adherence to one or another model, Moscow and Washington could not stand idly by. Leaders such as John F. Kennedy, Lyndon Johnson, Leonid Brezhnev, and even Mikhail Gorbachev were passionate believers in their respective systems and their capacity for global, not just domestic, change. As historian Odd Arne Westad put it, "Some of the extraordinary brutality of Cold War interventions—such as those in Vietnam or Afghanistan—can only be explained by Soviet and American identification with the people they sought to defend."[27] This is an observation that goes a long way toward explaining both Soviet and US behavior during the Afghan conflict itself—though again, it does not contain the whole truth. American identification with the *mujahadeen* probably outlasted Soviet identification with the Afghan communists or the Afghan people. But Soviet leaders never saw Afghanistan in isolation from the rest of their commitments and their relations with the Third World, and for this reason, among others, they felt they could not tolerate a loss there.

Cold War superpower interventions in the Third World are associated with a particular kind of counterinsurgency that follows what is now a familiar narrative arc: superpower army invades, find itself inadequate to the task of fighting agile guerillas, adjusts, wins battles, but ultimately withdraws. Another motif is the modernization project as counterinsurgency strategy and instrument from the Cold War toolbox. It certainly links the Soviet intervention in Afghanistan and the American war in Vietnam. As Michael Latham reminds us, Lyndon Johnson once hoped to create a "TVA [Tennessee Valley Authority] on the Mekong." The purpose of nation building in the counterinsurgency context was to give the South Vietnamese government more legitimacy: "Throughout the bulk of the war, US officials clung to the hope that they could create a South Vietnamese government with sufficient political legitimacy to win the war."[28] Both American and Soviet leaders hoped that their modernization projects would demonstrate to the local population that there were benefits to accepting the rule of their respective clients.

How and why the Brezhnev leadership decided to invade Afghanistan turned out to be of interest not only to scholars but to the Soviet government itself. Soon after the withdrawal in February 1989, a commission was set up to review this question. As a result, numerous documents were declassified and some were even published in the late Soviet era. Both Russian and Western scholars have made use of these documents to reevaluate the earlier conclusions drawn by their colleagues. The result was a radically different understanding of the motivations and decision-making process behind the Soviet decision to invade. If some contemporary commentators saw the invasion as part of a planned Soviet expansion toward the Persian Gulf, archivally based accounts of the decision to invade have shown that in fact Soviet leaders were largely responding to events in 1979.[29]

The change in our understanding of the decision to invade highlights the importance of archival research, combined with careful oral history and the mining of primary sources such as memoirs, for testing assumptions about how decisions were made in the Soviet Union. The story of the Soviet withdrawal from Afghanistan does not fit neatly into the existing interpretations of the changes in Moscow's foreign policy in the 1980s. I reject the view of contemporary Sovietologists who saw the war as a means of permanently extending Soviet influence and of making Afghanistan a Soviet republic in all but name. Such a view not only misinterprets Soviet leaders' decision making—it also ignores the agency of Afghan politicians, who sought extensive Soviet involvement in Afghan politics and economy and tried to delay the Soviet withdrawal. My interpretation also differs from that of writers who focus on the battle between New Thinkers and Old Thinkers within Moscow's elite. While New Thinking was undoubtedly an important paradigm in Soviet foreign policy during the Gorbachev period, the debate between New Thinkers and Old Thinkers leaves out key aspects of the story. For example, the military was not opposed to the withdrawal; and in policy debates on Afghanistan, senior officers took positions closer to Gorbachev's most reformist-minded aides, while reformists like Eduard Shevardnadze pursued a much harder line. Finally, what all previous accounts and interpretations have overlooked is the broad array of actors, beyond the top decision makers, who affected the timing of the withdrawal, its execution, and the diplomatic efforts to find a resolution to the conflict.

In the chapters that follow, we will see that four key paradigms shaped the involvement of the USSR in Afghanistan and determined the slow pace of its disengagement from the conflict. First, by the 1970s, aid to the Third World had become a key component of the Soviet bloc's legitimacy as a superpower. If the USSR's position in Europe was justified by its defeat of fascism, frequently recounted in movies, books, monuments, and demonstrations, then its position as a world power was justified by its defense of emerging states against encroaching neo-imperialism. The possible effects of a defeat in Afghanistan on the Soviet Union's reputation was a concern not only for Old Thinkers like Brezhnev and Andropov, but even for many in the reformist group that took over after 1985, not least Gorbachev himself.

Second, Moscow's extended presence in Afghanistan issued from its belief in what it could do to transform the country on behalf of its client regime. Even though leaders in Moscow recognized that the Soviet example was inappropriate for a country as underdeveloped as Afghanistan, they believed that they could go a long way toward stabilizing its client government in Kabul through a mixture of political tutelage and modernization programs. Thousands of Soviet advisers were sent to help the People's Democratic Party of Afghanistan improve its organizational work and gain support in the countryside. As we will see, these advisers sometimes did more harm than good. Even after most of them were withdrawn, in 1986, Moscow continued to try reform programs (such as the Policy of National Reconciliation) to stabilize the Afghan government. As with military aid, Moscow's presence was prolonged by a desire to give its programs a chance to work.

Third, despite a general consensus at the top of the Soviet hierarchy on Moscow's goals in Afghanistan, there was often little coordination among the various groups working in Kabul. The sharpest conflict was between the Soviet military and the KGB. Officers of the two security forces even tended to take sides in the internal PDPA split—the military supporting the Khalq and the KGB supporting the Parcham. These disagreements allowed Afghan communists to play the various sides against one another, and even to develop a lobby for their views in Moscow.

Finally, the conflict was prolonged by the high level of Soviet-US tensions in the 1980s. Although Pakistan armed and trained the *mujahadeen* opposition in Afghanistan, it was US money and resources that kept the *jihad* going, with help from Saudi Arabia, China, and several other countries. Soviet leaders believed that a settlement on Afghanistan would be possible only if the United States agreed to stop supporting the *mujahadeen*. At the same time, Moscow was cautious about opening a dialogue with the United States, fearing that doing so would be an admission that the invasion had been a mistake and that it would lose the freedom to act as it saw fit in Afghanistan. As we will see, Soviet-US tensions hindered Andropov's efforts to end the conflict in 1983, and they also made it more difficult for Gorbachev to seek a diplomatic solution during his first years in power.

This is the story of leaders coming to terms with painful truths. From 1980 to 1985, Soviet efforts in Afghanistan had three main objectives: first, to fight the opposition while simultaneously training and developing the Afghan army; second, to strengthen the regime in Kabul and make it more attractive to the population; third, to conduct diplomacy that would help the Kabul regime gain recognition and that would cut off foreign aid to the opposition. What was initially envisioned as a quick intervention became a long-term occupation. Various voices in Moscow tried to make their opposition to the invasion known to top leaders, and Soviet leaders eventually came to accept the need for UN diplomacy in resolving the Afghan conflict, and pushed their Afghan clients to do the same. Contrary to the argument advanced by some Western authors, the military costs of the war did not serve as a motivation for the withdrawal. Similarly, the social costs of the war were never sufficient to push Soviet leaders to seek a quick withdrawal. Indeed, the desire to withdraw, which was pushed most forcefully by Gorbachev's reformist New Thinkers, was balanced by concerns for how a withdrawal would be seen by other Third World states. This "correlation of forces" explains Gorbachev's decision to seek withdrawal gradually, rather than ending the war in 1985.

By 1986, though, Moscow was exploring ways to end the conflict on a number of diplomatic fronts. At the same time, it undertook a major reform of its efforts within Afghanistan and pushed the Afghan government toward the Policy of National Reconciliation. The two were related, as Moscow's diplomatic efforts at the time focused not only on dealing with countries such as the United States and Pakistan, but also on helping Kabul reach out to opposition leaders. In the end, the failure of these efforts to bring significant results led the USSR to seek a withdrawal without waiting for major improvements in Afghan politics by appealing directly to the United States.

Determined to withdraw Soviet troops and improve relations with the West, Gorbachev was ultimately willing to sacrifice the long-standing Soviet position on stopping arms supplies to the Afghan resistance in the hopes that improving relations with the United States would lead to a settlement in Afghanistan and elsewhere in the Third World. Ultimately, however, his misjudgment of American politics and decision making, as well as his inability to renege on traditional Soviet commitments, meant that a Soviet withdrawal did not lead to a resolution of the Afghan conflict.

The failure to coordinate the efforts of the various Soviet agencies active in Afghanistan was felt acutely during the withdrawal period, when the Najibullah regime seemed on the verge of a crisis. The military and the KGB clashed over how far to go in seeking an accommodation with the Tajik commander Ahmad Shah Massoud, halting the withdrawal, and managing the military operations. Even after the withdrawal, the declining Soviet regime continued to support its client in Afghanistan until the USSR's collapse in 1991. Arms, economic aid, and military advisers continued to prop up the Najibullah regime as long as the Soviet Union could provide them. Moscow was unable to bring about a reconciliation in Afghanistan, but Gorbachev did succeed in limiting the domestic fallout from the war by focusing the public's attention on the errors of Brezhnev, Andropov, and others who had originally decided on intervention.

1

THE RELUCTANT INTERVENTION

Although the main focus of this book is the Gorbachev period, one cannot understand the context in which he and his colleagues made decisions without considering the first five years of the war. In this period, Soviet policy was still made by the "old guard"—people like the CC CPSU General Secretary Leonid Brezhnev, the KGB chairman (and later General Secretary) Yurii Andropov, and the long-serving foreign minister Andrei Gromyko—while rising stars like Mikhail Gorbachev and Eduard Shevardnadze largely watched from the wings when it came to key issues of foreign policy. Soviet leaders began to seek paths beyond their military activities and beyond their support of the Afghan communist regime to settle the worsening situation in and around the fledgling socialist state. At the same time, concerns about maintaining prestige, as well as the deteriorating bilateral relationship with the United States, the USSR's chief Cold War rival, meant that the Soviet leadership moved slowly and often reluctantly to bring in outside help, such as that of the United Nations.

During the first months of 1980, Soviet leaders decided on an open-ended commitment of Soviet troops in support of the PDPA regime, complementing their military campaign with an influx of aid and political advisers. The key principles of Soviet policy in Afghanistan on the military, political, and diplomatic fronts were largely developed during this period. By 1985, these were: (1) to fight the opposition while simultaneously training and developing the Afghan army; (2) to strengthen the regime in Kabul and make it more attractive to the population through economic aid and political tutelage; and (3) to conduct diplomacy that would help the Kabul regime gain recognition and that would end foreign aid to the opposition. Under Gorbachev's leadership, the Soviet Union was slow to depart from these principles.

Soviet leaders could bide their time because in every sense the Afghan intervention was a limited war. The Soviet military adjusted to the demands of a prolonged counterinsurgency campaign that it had not planned for, and military losses remained at a tolerable level throughout. Similarly, the war's effect on Soviet society during the period in question was still restricted and did not force the country's leadership to take drastic measures. While knowledge about the war and dissatisfaction with the Soviet involvement grew as the 1980s progressed, this did not translate into public pressure on Gorbachev or his colleagues to end the war immediately. Soviet Muslims did not become "infected" with a desire to wage *jihad* on the Soviet state, despite the predictions of some Western experts. The war remained limited both in its effects within the Soviet Union and in terms of the military and economic resources it required—a crucial point in the understanding of why Gorbachev did not bring the troops home quickly when he came to power in 1985.

The Reluctant Intervention

The story behind the intervention has been known for some time, but it is worth recounting here. Not only does it help to put the rest of the story in context; it also shows certain patterns in Soviet decision making that affected the conduct of the war, the relationship with the Afghan communists, and the way the situation in Afghanistan was viewed in relation to Soviet foreign relations elsewhere.

What became known as the Saur Revolution, which brought the PDPA to power, was really a reaction to Daoud's purges against the left beginning in 1977. In April 1978, a senior Parcham member, Mir Akbar Khaibar, was assassinated, and his funeral turned into an antigovernment demonstration. Daoud decided to arrest the communist leadership, including Taraki. Hafizullah Amin, a fellow Khalqi, was merely placed under house arrest. This proved a fatal oversight for Daoud. Amin was able to use his connections in the military to launch a coup. In the ensuing fighting, Daoud was killed. Colonel Abdul Kadir, a leading mutineer, helped to establish a revolutionary council which in turn elected Taraki as the prime minister and president.

Afghanistan's new leaders quickly showed they had little interest in governing along with officials left over from the monarchy and Daoud's republic. Seraya Baha, a former Parcham activist who had left the party but stayed friendly with its leaders, including Mir Akbar Khaibar, was fired from her job at the planning ministry within days of the revolution.[1] (Her marriage to the brother of Najibullah, a rising Parcham star who would one day become president, did not protect her.) Seraya Baha's boss, the government minister Abdul Karimi, was accused of being a CIA spy and fired as well, replaced by Sultan Ali Keshtmand. Daoud supporters were purged from the government at all levels. Nor did power unite the factions of the PDPA—indeed, it only polarized them further. The Khalqis were eager to move ahead quickly with their modernization and redistribution program, and they were intent on concentrating power in their own hands. A reign of terror was unleashed on two fronts: against traditionalist elements, especially members of the clergy, followers of the Muslim Brotherhood or of Ayatollah Khomeini; and, simultaneously, against the "enemy within," primarily Parchamists. Many were thrown in jail, while others, like Babrak Karmal, were sent into diplomatic exile. (Karmal served as ambassador to Czechoslovakia during this period.)[2] Only three months after taking over the planning ministry, Sultan Ali Keshtmand was himself arrested for allegedly plotting against Taraki.

KGB officers in Afghanistan were alarmed by the extent of Taraki's terror and the rapid pace of his attempts to transform Afghan society. They feared that his reforms and repressions would undermine the young government and throw the country into chaos. In July 1978, the Kabul residency sent an appeal for a political intervention at the highest levels: "only the leadership of the CPSU can influence the wild [Khalq] opportunists and force them to change their attitude toward the Parcham group."[3] Indeed, Boris Ponomarev, head of the International Department of the Central Committee, traveled to Kabul at the end of September to press Taraki to "stop the mass repressions, which have taken on increasing proportions following the revolution in Afghanistan, including repressions against the Parcham."[4]

> Emphasis was also placed on the importance of creating and strengthening the party throughout all of the country's territories, on the adoption of prompt measures to normalize the activities of party organs from top to bottom, on organizing agencies of the people's government, and on focusing increased attention on economic problems. The people must experience concrete results of the revolution in their own lives. That is why the improvement of people's lives should be the primary focus of the new government.[5]

Moscow's concerns did not stop the two countries from signing a Treaty of Friendship on December 5, 1978. By then, a number of Soviet advisers were serving in the government, the party, and the military. The latter were advising their tutees in military operations against the emerging rebel groups.[6]

The first great test of Soviet commitment, however, came in March 1979 with the uprising in Herat. This ancient and largely Tajik city, located in western Afghanistan, erupted in revolt on March 15. A mutiny led by mid-level officers joined with a mass uprising of the city's residents.[7] Afghan officials, Soviet advisers, and their families all fell victim to the mob violence that overtook the city. The Afghan leadership lost its nerve, believing that its own military would be unable to deal with the situation. They called on Moscow to send Soviet troops and planes to quash the uprising.

The Politburo met several times over the following days to discuss the situation. At first, key foreign-policy makers within the Politburo supported intervention. Dmitrii Ustinov, the minister of defense, Andrei Gromyko, the minister of foreign affairs, and Yurii Andropov, the KGB chief, spoke in favor of armed intervention at a Politburo meeting on March 17, arguing that the risks of engaging Soviet troops outweighed those of losing Afghanistan. Afghanistan was too important, Gromyko insisted, to let it fall into hostile hands: "If we lose Afghanistan now and it turns against the Soviet Union, this will result in a sharp setback to our foreign policy."[8]

By the next time the Politburo met, however, the situation had changed: intervention was seen as inadvisable both in view of the situation in Afghanistan and because of the threat it would pose to détente. With the Carter-Brezhnev summit scheduled for that June in Vienna and the expected culmination of negotiations on SALT II (the follow-up to the first Strategic Arms Limitation Treaty), there was too much to lose. Ustinov and Andropov both realized that the Soviet army would end up fighting on behalf of the Afghan army. Apparently, Brezhnev's foreign-policy adviser, Andrei Aleksandrov-Agentov, played a key role, pushing his boss to override Ustinov, Gromyko, and Andropov's enthusiastic support for intervention.[9] These three leaders, along with international department chief Boris Ponomarev, would form the Afghanistan Commission of the Politburo, and their dominance over decision making in this area only increased.[10]

The Afghan army was able to pacify Herat, but the country was moving away from normalcy, the government still unable to find its footing. The PDPA was further from unity than it had ever been, and a conflict was growing between its two top leaders, Taraki and Amin. Moscow instructed its officials on the ground to take an active role in trying to resolve it, but with little success.[11] As would happen many times throughout the intervention, each side in the intra-PDPA contest had its supporters among the Soviet advisers. In this case, some of the military advisers had been impressed by Amin's role in putting down the Herat revolt in March. Taraki, however, had Brezhnev's support. During a visit by Taraki to Moscow in September 1979, Brezhnev and Andropov warned him that Amin was planning to oust him. When Taraki returned to Afghanistan, he tried to act on this information by having Amin killed, possibly with KGB help. In the event, the attack failed; Amin escaped unharmed and had Taraki arrested. The poet-turned-revolutionary-leader was strangled several weeks later on Amin's orders.[12]

Taraki's arrest and murder seem to have started the final sequence of events that led to intervention. At first, Soviet leaders tried to make the best of the situation, instructing their officials in Moscow to accept Amin's consolidation of power as a fait accompli while working to minimize repression against supporters of Taraki.[13] Brezhnev seemed resigned yet cautiously optimistic at a Politburo meeting on September 20: "We should assume that Soviet-Afghan relations will not sustain some sort of major changes, and, it seems, will continue in their previous course. Amin will be pushed toward this by the current situation and by the difficulties which the Afghan government will face for a long time to come. Afghanistan will continue to be interested in receiving from the USSR military, economic, and other aid, and possibly even in increased amounts."[14] Yet Amin was proving an increasingly difficult partner. Soon after having Taraki killed, he in effect expelled the Soviet ambassador, Aleksandr Puzanov. Nor did he adhere to Soviet requests to refrain from repression against fellow PDPA members. A memorandum from the "Afghan commission" dated October 29 noted that in light of this, Moscow ought to continue working with Amin, but also remain vigilant for a "turn by H. Amin in an anti-Soviet direction."[15]

There is evidence that at some point in October or November, Ustinov and Andropov began reconsidering their earlier agreement to hold off on armed intervention. Amin's erratic behavior, including reported secret meetings with US officials, was part of the reason; a worsening international situation was another. The Islamic revolution in Iran made senior Soviet planners wonder if the US would now look at Afghanistan as a new base for its forces in the Persian Gulf. The Carter administration's decision to move naval forces into the area in the fall of 1979 only fueled Soviet suspicions.[16]

Ustinov and Andropov now formed the chief pro-intervention lobby, and they apparently convinced Gromyko, as well as Aleksandrov-Agentov, to support their arguments. In early December, Andropov wrote a personal memo to Brezhnev laying out the case for intervention. It highlighted Amin's untrustworthiness and the possibility that he might go over to the West.

> The situation in the government, in the army, and in the state apparatus is aggravated. They are practically disorganized as a result of mass repressions carried out by Amin. At the same time, we have been receiving information about Amin's behind-the-scenes activities which might mean his political reorientation to the West. He keeps his contacts with the American chargé d'affaires secret from us. . . . In closed meetings, he attacks Soviet policy and the actions of our specialists. . . . Now there is no guarantee that Amin, in order to secure his personal power, would not turn over to the West.[17]

Andropov also offered a solution. The Parchamists whom Amin and Taraki had expelled could be brought back into the country and form the core of a new government. A limited military force, consisting of two battalions already stationed in Kabul, would be needed, but a larger group would be kept along the border "just for emergencies." Such an operation, Andropov concluded, "would allow us to solve the question of defending the achievements of the April revolution, resurrecting the Leninist principles of state and party building in the Afghan leadership, and strengthening our positions in that country."[18]

Ustinov and Andropov met with Brezhnev on December 8, 1979, to advance the case for intervention. Their arguments pointed out that, realigned toward the West, Afghanistan could well become the staging area for missiles directed at the Soviet Union.[19] Once Brezhnev's support had been secured, only the formal matter of a Politburo resolution remained. On December 12, the Politburo met for a brief session and approved a handwritten resolution entitled "Concerning the Situation in A."[20]

When Soviet leaders approved the intervention, they did not envisage fighting a war on behalf of the PDPA. Indeed, Andropov preferred that only a very limited number of troops be committed in support of the operation to remove Amin.[21] Ustinov, however, insisted on a larger contingent, comprising 75,000 troops. Their purpose was to boost morale and take a defensive posture in Kabul, as well as in some provincial capitals. By protecting Afghan military installations, they would free the military of the DRA (Democratic Republic of Afghanistan) to handle any uprisings. The removal of Amin would be handled by an elite brigade.[22]

The decision to intervene was not without its opponents. They included senior military officers who tried to make their case to Ustinov in the weeks leading up to the intervention. According to testimony from several senior General Staff officials, they appealed to Ustinov in particular not to support the introduction of troops.[23] The last such effort took place on December 10, two days before the government made the final decision to intervene. Nikolai Ogarkov spoke on behalf of the General Staff, setting out to Brezhnev, Gromyko, Andropov, and Ustinov the reasons the Soviet Union should not send in troops. According to General Valentin Varennikov—Ogarkov's deputy, who would go on to lead the operational group in Afghanistan—his boss made the following points:

> First, that the Afghans should deal with their internal affairs themselves, and we should only give assistance; second, that the public would not understand us—neither the American people, nor the Soviet people, nor the world in general—if we introduced those troops; third, that our troops did not know the specific circumstances of Afghanistan very well—the tribal relations, Islam, and various other things would put our troops in a very difficult situation. And he made some other arguments.[24]

It is not surprising that the military was wary of an intervention in Afghanistan. This was clearly not going to be like other Soviet involvements in the Third World, where Soviet officers were fulfilling their "internationalist duty" by advising and training local forces, a prestigious assignment. Here they would be commanding Soviet troops who, if things went wrong, could well end up fighting Afghan insurgents. Their concerns were partly practical: as Ogarkov pointed out, the Soviet army was not trained to fight guerrilla wars in a country like Afghanistan—its main enemies were NATO and China, and its planning and its training were geared toward battle in Central Europe and East Asia, or in the skies above. But there was also a psychological factor. Many of the senior officers had fought in World War II—Varennikov had been a captain and had participated in the taking of Berlin in 1945. Their sacrifices and their victory were frequently celebrated and had met with genuine gratitude from the Soviet people, and they could hold their heads high. Younger officers had been brought up with the memory of that generation's great victory, its successful defense of the motherland, and they wanted to carry that flag. They understood very well that in Afghanistan the war would be messier, and that any failures there were likely to be blamed on them.

Ogarkov's arguments failed to impress the Politburo members, who had already decided on intervention. Although in March similar arguments had persuaded them to reject military intervention as an option, now they seemingly saw no other way to handle the situation. The failure of the US Congress to ratify SALT II in the summer of 1979, which seemed to signal an American turn away from détente, was one reason. The US decision to deploy Pershing missiles in Europe was another.[25] The murder of Taraki by his rival Hafizullah Amin, despite Brezhnev's pledge of support, helped convince Brezhnev that the latter had to be removed from power.[26] He would later complain to French president Valéry Giscard d'Estaing: "President Taraki was my friend. He came to see me in September; and just after he returned, Amin had him assassinated. That is a provocation. I could not pardon it." Growing suspicion that Amin might be considering a turn toward the United States contributed to Brezhnev's hostility.

Several conclusions can be drawn about the decision making that led to the invasion. First, the invasion was the result of a decision reached by several key foreign-policy figures within the Politburo, not by the Politburo as a whole. This was characteristic of decision making in the late Brezhnev era. With Brezhnev himself ailing, foreign policy was dominated by three people: Andrei Gromyko, the foreign minister; Dmitrii Ustinov, the minister of defense; and Yurii Andropov, chairman of the KGB. At the same time, advisers such as Brezhnev's foreign-policy aide Andrei Aleksandrov-Agentov played key roles in shaping decisions.

Second, dissenting voices from within the Politburo, as well as from other ranks of the Soviet bureaucracy, were regularly silenced by this troika. According to Karen Brutents, an official in International Department, throughout the late fall of 1979 Aleksandrov-Agentov even pushed those preparing to advise against intervention to abandon their position.[27] Similarly, senior military officers who tried to object to the operation were told to mind their own business and "not teach the Politburo." Once Gromyko, Ustinov, and Andropov had come to an agreement among themselves and managed to secure Brezhnev's support, they were able to pressure other Politburo members and senior officials to accept their decision.[28]

Finally, those who supported the decision to invade did so because they felt that the "loss" of Afghanistan would be unacceptable and a blow to Soviet prestige. At the same time, it did not mean that they had completely abandoned détente. Leonid Brezhnev's commitment to détente was very strong, as was that of Gromyko and Andropov. Nevertheless, it did not override other concerns these men shared as leaders of a great power that had client states around the globe and that was in an ongoing contest with the United States and China for influence, particularly in the Third World. The fact that détente hit a low point after the rejection of the SALT II treaty by the US Congress in the summer of 1979 served as a catalyst for supporters of the invasion. The "loss" of Afghanistan would be particularly embarrassing at a moment when the Soviet Union's main adversary seemed to be abandoning détente. Later efforts to extricate the Soviet troops from Afghanistan would often move with the ebb and flow of the USSR's relationship with the United States.

From Intervention to War

The purpose of the initial Soviet intervention in December 1979 was limited: change the leadership, then garrison the cities and protect key bases, so that the Afghan military would be free to quell disorder and fight any insurgency. Its planners expected that Soviet troops would be able to return home within several months. The long-time Soviet ambassador in Washington, Anatolii Dobrynin, recalled that when he brought up his concerns about the damage the invasion would do to Soviet-American relations, his boss, foreign minister Andrei Gromyko, replied: "We'll do everything we need to in a month and then get out."[29] Brezhnev confirmed this, saying the troops would be out within several months.[30] The situation in Afghanistan, however, turned out to be worse than Soviet leaders had anticipated. The Karmal regime could not quickly establish its control in the countryside, and the Afghan army was in bad shape, demoralized by the purges and divided over the country's course. Within several months, the Soviet leaders who dominated foreign policy decided that the stabilization of the PDPA regime required a long-term commitment of Soviet troops. Rather than just providing training and some security, these troops would engage the Kabul regime's enemies directly.

Western scholars and analysts have suggested that Soviet leaders were suffering from a "Czechoslovakia syndrome" when they intervened in Afghanistan. In 1968, Soviet troops had managed to restore a pliable conservative regime in Czechoslovakia, after several months of worry over the nation's experiment with a more liberal form of communism. Although the invasion had been condemned by Western countries and even by some Soviet citizens, the mere presence of Soviet arms had settled the situation, and a sort of calm quickly returned. There is nothing in the documents to suggest that Soviet leaders were thinking of Czechoslovakia when they considered sending troops into Afghanistan, although there were certain similarities in the way the actual invasion was planned.[31] Indeed, in earlier discussions of a possible intervention, Soviet leaders clearly expressed their concern that such an intervention would lead to Soviet troops fighting the Afghans directly. In the end, this was exactly what happened.

The goal of the invasion was to secure infrastructure, free up the Afghan army to conduct raids and operations, and enable the new government to function. Soviet leaders did not envision their army being directly involved in battle after the initial invasion—they were there just to prop up the military of the Democratic Republic of Afghanistan.[32] Almost immediately, however, the Limited Contingent of Soviet Troops (really the 40th Army, commonly referred to by the Russian acronym OKSV) was faced with a situation that foreshadowed the difficulties of working with the Afghan army. In early January 1980, the 4th Artillery Regiment of the DRA army, based in the northern settlement of Nahrin, mutinied. Since it was suspected that Soviet advisers had been murdered, Limited Contingent troops were sent in to quell the insurgency.[33] David Gai and Vladimir Snegirev, both of whom traveled to Afghanistan on numerous occasions during the war and interviewed Soviet and Afghan participants, write that over one hundred mutinous soldiers were killed.[34]

Nevertheless, the new Parchamist leadership had reason to hope, in those first few weeks after the invasion, that they could establish control over the county. Amin's short reign was bloody, and his removal was welcomed. A KGB official working in Afghanistan recalled that Soviet soldiers were greeted warmly and told, "You have done a great deed by removing the bloody Hafizullah Amin," but were warned to go back to their homeland quickly.[35] Naturally there was genuine relief that Amin was gone and that a more conciliatory leader had replaced him. Soviet leaders, hoping to avoid a repeat of Amin's brutal reign, urged Karmal to be much more lenient with the political opposition, even stopping him from severely punishing former Amin supporters. The removal of Amin was welcomed not only in Kabul but even in the provinces, leading many rebels to lay down their arms.[36]

It soon proved impossible, however, to keep Soviet soldiers out of the fight. Following the use of the OKSV soldiers to put down the mutiny at the beginning of January, Soviet troops were drawn into skirmishes with increasing frequency. Officers and soldiers of the Red Army noticed anti-Soviet propaganda spreading quickly throughout the towns and villages; and by the end of the month, it seemed to some of them that the only pro-Soviet Afghans were those who worked for the PDPA.[37] To the supporters of the intervention, this was all the more reason for Soviet troops to continue, and even expand, their mission. The new Afghan leaders just needed time to reassert their authority—and besides, the international reaction was not all that bad. Andrei Gromyko, the foreign minister, told his colleagues that world public opinion was divided and not at all solidly in the US camp. Brazil, Argentina, and Canada, for example, did not want to follow the US lead of stopping grain sales to the Soviet Union. Yurii Andropov noted the major effort by Babrak Karmal to create unity within the party and to reach out to tribes and certain members of the clergy. Over the previous few weeks, he pointed out, the government had once again started to take on solid shape, acquiring "all the necessary organs of party and state leadership."[38] The small outbreaks of violence, as well as the anti-Soviet and antigovernment propaganda described by officers, did not seem to worry Soviet leaders greatly.

Soon after the invasion, the interventionists in the Politburo had to face jittery colleagues who wondered when the troops could be brought home. The introduction of Soviet troops had raised the stakes, however: they could not come home until the situation there was markedly better —with the Kabul regime far more secure—than it had been before they went in. A fragment of a Politburo discussion at the end of January 1980, following a trip by Andropov to Kabul, makes it clear that someone had suggested the possibility of withdrawing troops. Ustinov and Gromyko spoke against this. The former suggested it would take at least a year to pacify the opposition; the latter seemed even more pessimistic. He pointed out that it would be dangerous to leave before there was some written agreement between Afghanistan and the countries supplying the opposition with arms. "We will never have a complete guarantee, I think, that no hostile country will ever again attack Afghanistan. That is why we need to provide for Afghanistan's complete security."[39] While accepting the need to begin working through diplomatic channels to help secure Afghanistan's position, Gromyko was still arguing for an essentially open-ended commitment to maintain the regime's position through the use of Soviet troops.

As protests and small attacks on the regime became more widespread, supporters of the intervention became increasingly amenable to the direct use of Soviet troops to attack the opposition. Karmal, like the Khalqis he replaced, hoped that Soviet troops would take a more active role in helping him quash the armed opposition. According to Liakhovskii, both Marshal Sergei Sokolov and General Sergei Akhromeev, the two top-ranking soldiers in Afghanistan, had been able to avoid the commitments Karmal requested in the first months of the intervention. Increasing hostility to the presence of Soviet troops and the DRA government, however, convinced Moscow that Soviet troops would have to engage the enemy directly. On February 20, a major protest broke out in Kabul. Hasan Kakar, at the time a professor of history at Kabul University, wrote that it was the largest protest Kabul had ever seen, involving crowds of thousands in different parts of the city.[40] On February 23, opposition militants had attacked the USSR's embassy in Kabul, as well as several Soviet encampments.[41] The event seems to have unnerved local Soviet representatives and the DRA leadership, who sent urgent requests to Moscow that Soviet troops be allowed to "liquidate the enemy."[42] A directive then arrived from Moscow, ordering the 40th Army to conduct joint operations with the army of the DRA.[43]

By March, the Soviet army was involved in full-scale operations, repelling guerrillas who were advancing on Asadabad, the capital of Kunar Province. The incident foreshadowed a pattern in several ways. Soviet troops were called in to help in an area where the Afghan army had at first seemed to be in control. They were able to beat back the guerillas through intense shelling, which, however, prompted an exodus of civilians into Pakistan.[44] When the Soviet army left, the guerillas resumed their attack on the DRA forces. As one observer put it, "[Afghan troops] seemed confident only when they were near Soviet troops."[45] The more Soviet troops took part in battles, the more the Afghan army seemed to limit itself to "mopping-up" operations.

Moscow's new goal in Afghanistan was to establish the security necessary for the Karmal regime to take root and be able to withstand military and political challenges. At the same time, Soviet leaders had little confidence that their Afghan protégés, either in the party or in the military, could stand on their own. As in January, oral and written statements by members of the Afghanistan commission noted that considerable progress had been made by the Karmal government in restoring the authority of the state, but that it was too early to think about the withdrawal of troops. In March, they told the Politburo that although the government was taking proper measures with regard to its position domestically as well as internationally, the process was "moving slowly." At the same time, their memorandum said, "the fighting ability of the Afghan troops remains low."[46]

The scarcity of documentation makes it difficult to determine the overall mood in the Politburo on the question of the OKSV and its role. However, it seems that at least some members expressed concern about the consequences of keeping troops in Afghanistan. Liakhovskii cites a document from late February 1980 suggesting that Brezhnev brought up the question of a withdrawal, but that the possibility was rejected by Ustinov and Andropov.[47] It may well be that Brezhnev was unhappy with the possibility of an indefinite presence of Soviet troops in Afghanistan, although earlier that month he had himself brought up the possibility of sending more troops.[48] It would certainly be consistent with his oft-demonstrated commitment to détente. In any event, he supported the members of the Afghanistan commission when they argued for putting off any withdrawal. In all likelihood, he was genuinely upset by the possibility of a long-term commitment of Soviet troops, but did not know how to proceed and thus relied on his colleagues and advisers to direct policy. Still, it was not just Brezhnev who wanted to see Soviet troops come home as soon as possible, as is clear from the extremely vigorous defense of their continued presence in Afghanistan by supporters of the intervention.

The architects and defenders of the intervention believed that the Soviet Union had made the right decision in intervening and was making the right sort of investment in the country. In the Politburo, they focused on calming the nerves of jittery colleagues who sought to limit the presence of Soviet troops. On April 7, they presented a memorandum that listed the benefits the Soviet invasion had brought to Afghanistan, as well as to the Soviet Union's security.[49] The Soviet Union had invested too much in Afghanistan to withdraw prematurely. For the time being, Soviet troops would have to play a leading role in defending the regime, and a Soviet military presence would be required for a long time: "The successful resolution of internal problems and the strengthening of the new order in Afghanistan will take significant effort and time, during the course of which Soviet troops will continue to be the key stabilizing factor," the Afghanistan Commission had written in March.[50] In fact, the 40th Army was assuming all the responsibilities of a national army, as the April memorandum made clear:

> Our troops in Afghanistan will have to continue fulfilling the task of defending the revolutionary order of the DRA, defending the borders of the country, and providing security in key centers as well as transportation links. . . . Only with the stabilization of the internal situation in Afghanistan, as well as the improvement of conditions around it, would it be possible, at the request of the DRA leadership, to consider the question of a gradual withdrawal of Soviet troops from the DRA.[51]

Within several months of the invasion, any hope of a quick turnaround evaporated. Originally, Soviet troops had entered to save a revolutionary government from an erratic leader and to make sure an ally did not go over to the US camp. Now they were there to make sure a new government installed through that intervention could stay in power.

The Afghanistan Commission developed the idea for the intervention and was the key policy-making body in the first years of the invasion. The public knew very little, and the CPSU was not involved in any decision making. With the realization that Soviet troops would have to stay in Afghanistan for a longer period of time, however, it was necessary to go through the formality of securing party endorsement for Soviet policy in Afghanistan. At a special plenum convened in June 1980, Gromyko delivered a speech defending Soviet policy in Afghanistan. The Soviet Union, he said, would not apologize for sending in troops; indeed, "the ones who should be apologizing are those who are behind the aggression against Afghanistan, who carried out the criminal plans with regard to this country." Furthermore, Gromyko said, echoing an earlier statement by Brezhnev, it was necessary to keep the troops there.[52] The plenum voted to "fully approve" the actions taken by the leadership.[53] Both the initial invasion and the continuing presence of the 40th Army in Afghanistan now had the official support of the party.

It is significant that the Afghanistan Commission consisted of the Soviet Union's most senior politicians, all of them close to Brezhnev. There was no more powerful constellation of personalities than Gromyko, Andropov, Ustinov, and Ponomarev in the CPSU of the early 1980s.[54] They represented, collectively, the Foreign Ministry, the KGB, the Ministry of Defense, and the International Department. With Brezhnev on their side, they also represented the party. In experience and in formal position, each one of them individually was among the highest-ranking members of the Politburo. Taken together, they also represented the chief institutions responsible for the conduct of foreign affairs.

In the first six months after the invasion, these leaders came to accept the need for an open presence of Soviet troops in Afghanistan. They did this not from any desire to "colonize" Afghanistan, but because they did not believe that the Karmal government was ready to stand on its own. The growing insurgency, rather than discouraging the extended presence of Soviet troops, convinced Moscow that their use was necessary and appropriate. From the end of February 1980, Soviet troops would become the DRA's main fighting force.

Nation Building in Afghanistan

Moscow's broader strategy in the early years of the war aimed at uniting the PDPA; giving it greater legitimacy through the use of traditional Afghan institutions, including tribal councils and the clergy; and making the regime more attractive through infrastructure programs and other aid. The so-called Policy of National Reconciliation, launched in January 1987 with enthusiastic support from Gorbachev and discussed in Chapter 4, was largely a reformulation of the policy described below. At the same time, the domestic and international situation that had developed as a result of Amin's repressive rule and the Soviet invasion meant that the new Karmal government was greatly dependent on Soviet aid, trade, and specialists.

With Soviet encouragement, the new Afghan government took a number of steps to slow down or even reverse some of the reforms of the Taraki and Amin period. Land confiscated from "middle peasants" and even some larger landholders was returned. The legal limit on landownership was increased as well, and refugees who had abandoned their land were promised restitution. Mosque land was no longer in danger of confiscation. Parcham leaders saw land reform as crucial to the success of the revolution, but took care to avoid antagonizing social groups who were not natural allies of their government. They recognized that the radical efforts of Taraki and Amin had damaged agricultural production and had turned whole sectors of the population against the government. The new leadership's more pragmatic politics extended beyond land reform to policies on women's education and participation in literacy courses.[55]

Contemporary western commentators interpreted the "Sovietization" of Afghanistan after the invasion as part of a broader plan to make Afghanistan a virtual republic of the USSR. They noted the growing share of Soviet exports and imports in Afghanistan's foreign trade, the ever-increasing number of Soviet specialists, and the extent to which Afghan government and enterprises were organized on Soviet models. In fact, this was due to the fact that the Karmal government had few friends outside the Soviet Union and its allies. The hostility of Pakistan and Iran to communist Afghanistan meant that trade with these natural (in terms of geographic proximity) partners was severely restricted. So was trade with other traditional partners such as India, where goods had to cross hostile territory.[56]

The USSR became not only Afghanistan's major trading partner, but also a clearinghouse for Afghan imports and exports.[57] Besides nonrepayable aid (which in terms of consumer goods alone amounted to 210 million rubles in 1986, for example), the USSR also provided Afghanistan with credits that were to be used for buying Soviet products and repaid with Afghan exports. Since demand for Afghan exports other than natural gas (such as rugs, wool, and dried fruits) was low, the DRA was never able to repay these credits; but the credits were generally extended or forgiven by Moscow.[58]

Trade and material aid were only some of the ways that Moscow tried to help the Karmal government. The Soviet leadership understood that in order to stabilize the country, Karmal would need to unify the party as well as convince the rest of the country to accept PDPA rule. At the end of January 1980, the Afghanistan Commission presented a plan of action which called for measures to spread the PDPA's influence into the countryside with the help of party activists and youth organizations. At the same time, it tried to take into account the specifics of Afghan power structures. The document called for efforts to reach out to tribal leaders, the use of jirgas (traditional tribal councils), and a "long-term plan for work with Muslim clergy."[59]

Before the PDPA could spread its influence into the countryside, Karmal would have to achieve a degree of unity within the party that had been elusive since its founding and that had been further undermined by Amin's purges. Moscow's concept of unity did not always match up with Karmal's, however. Soviet leaders wanted Karmal to form a government that included Khalqis, and helped to broker a deal between him and several Khalqi ministers in Moscow before bringing him to Afghanistan. Once in power, Karmal began to edge out Khalqis, even executing some of Amin's closest associates.[60] The only reason a full-scale purge did not take place was that Moscow made it very clear that this would be unacceptable. Soviet advisers pressed Karmal to stop the removal of Khalqis from party and administrative posts, and a formal CC CPSU request to this effect was sent to him sometime in January 1980.[61] Karmal, for his part, kept trying to gain a free hand, telling his Soviet interlocutors, "As long as you keep my hands bound and refuse to let me deal with the Khalq faction, there will be no unity in the PDPA and the government cannot become strong. . . . They tortured us and killed us. They still hate us! They are the enemies of the party!"[62]

To help stabilize the government and broaden its base of support, the Soviet Union sent thousands of technical specialists and political advisers.[63] Some were Soviet party workers sent to advise the party in Kabul and in the provinces. Many more were sent to factories, businesses, and universities. While technical specialists built ditches, operated mines, and extracted natural gas, political advisers wrote speeches on behalf of politicians and memoranda on behalf of ministers, and went out into the countryside to help Afghan communists reach out to the local population. Some took part in distributing Soviet aid in the countryside; others tried to coordinate pacification measures, helping to organize truces between rebel bands and the government. Soviet Embassy employees joked about the "limited contingent of Soviet party advisers in Afghanistan."[64]

Most party advisers did not have any sort of specialized training for the work they were about to undertake. In fact, "training" for a party adviser about to be sent to Afghanistan was apparently a one-week course dealing with the "political, military, and economic situation in that country," plus whatever additional reading on Afghan history or politics the soon-to-be adviser might pick up on his own. During the week-long course, instructors from the CC CPSU International Department emphasized the importance of the "internationalist mission" about to be undertaken and tried to inculcate a sense of optimism regarding the job.[65]

Yet other types of advisers working in Afghanistan were better prepared. Military advisers working with the political sections of Afghan units were largely in the same boat, preparation-wise, as their party counterparts. Some of the Foreign Ministry officials posted in Kabul were specialists on the region, if not on Afghanistan itself, and sometimes even had knowledge of Dari or Pashto.[66] Among the best-trained were advisers from the KGB and GRU (the Soviet national-security and foreign-intelligence agencies), who were sent to Afghanistan only after serious preparation, including "two years of Dari or Farsi, plus courses in Afghan history, economy, culture, customs and traditions, religion, and so forth."[67] Members of all the KGB missions that served in Kabul during the war had experience in the region, and three were academically trained orientalists.[68] On the whole, however, such well-trained advisers were hard to come by. The scale of the Soviet involvement meant that there was not enough time to prepare a well-trained cadre. Corners had to be cut, and thousands of advisers were sent virtually without preparation.

One unique resource the Soviets could draw on were advisers and interpreters recruited in Central Asia, particularly the Uzbek SSR and the Tajik SSR. In theory, at least, they had a deeper understanding of Afghan culture than their Slav colleagues. Most crucial, perhaps, was their knowledge of languages, particularly Farsi (of which Tajik is a dialect) and Uzbek. (Most of the interpreters who went to Afghanistan, whether Tajik, Uzbek, or Slav, were recruited from the major institutes of oriental studies or language programs at universities—for example, Tashkent.) The Soviets could also draw on the language specialists who were training at the various institutes and universities in language programs. The translators worked with party advisers, youth advisers, military advisers, and specialists.

The advisers were sent with general briefs and instructions from their respective organizations and from the party. Once they were "in country," however, these advisers had quite a bit of independence. While their work was theoretically coordinated by the chief political adviser (based in Kabul), in practice each group of advisers often acted independently. As a result, the groups were often working at cross-purposes. In some extreme cases, individual advisers took initiatives that were contradictory to Moscow's instructions.

The presence of Soviet troops and advisers seemed to cause paralysis among Afghan politicians. This may have been due to a sense that the Soviet advisers could do the job better, or it may have been a response to the generally imperial attitude adopted by some advisers. Often, Soviet advisers preferred to carry out a task themselves, rather than train their subordinates.[69] It was common practice, for example, to write speeches in Russian for translation into Dari and Pashto. This practice apparently included party documents and, later, the new constitution adopted under Najibullah.[70] A Soviet assessment of the PDPA from 1983 noted that even at the highest level of the party, there was a tendency to shy away from decision making. Karmal, Keshtmand, and the other members lacked initiative, the assessment said, and "turn to advisers not just for counsel, but also to transfer to them their own functions for the composition of working documents, instructions, and especially texts of reports and articles."[71] Karmal himself later confirmed that advisers had become ubiquitous in the party and bureaucracy; he admitted that many Afghans had largely stopped working, preferring to "lay all the burden and responsibility for practical work on the shoulders of the advisers."[72]

The "nation-building" effort failed on both the political and economic fronts. Economic aid often did not reach its intended destination, as items like trucks, tanks, cotton, and food products were either diverted and resold or fell prey to hijackers.[73] Infighting and lack of coordination among the Soviet advisers were constant problems. Everyone, complained a *Pravda* correspondent in a confidential report to the Central Committee written in 1981, felt that they should have the dominant voice. This bred chaos: "Conflicts between our advisers and representatives, and the lack of coordination in our positions and actions, greatly reduce the effectiveness of USSR aid to Afghanistan."[74]

Infighting and lack of coordination among advisers and other Soviet officials had numerous practical consequences that undermined the Soviet mission. In a number of cases, "liberation" of villages and successful efforts to win over rebel commanders floundered when some Soviet advisers or Afghan officials refused to cooperate. The party adviser Ilya Elvartynov recalls the difficulty he had in persuading Afghan ministers and Soviet officers to leave an armed detachment for his exposed team and in making sure that the established trade links with Soviet enterprises in Tajikistan actually worked the way they were supposed to.[75] Similarly, advisers who were involved in successful efforts to win over minor (but locally important) rebel commanders found their efforts undermined when local authorities refused to cooperate on the "payoff." For example, in 1985 KGB adviser Valerii Mitochkin, working with KhAD (the Afghan intelligence agency, dominated by the Parcham), was able to convince a commander—along with his family and three hundred fighters loyal to him—to make peace with the government. In exchange, all of the militants and their families would be given housing and employment, and were allowed to keep their weapons. But the local governor (a Khalqist) refused to cooperate in providing housing, while the local military commander insisted on seizing some of the weaponry to send back to Kabul. Furthermore, according to Mitochkin, the Soviet party adviser refused to cooperate and mediate between the governor on the one hand and Mitochkin and KhAD on the other.[76] Advisers elsewhere faced similar difficulties. Such problems damaged the reputation of advisers, Afghan authorities, and pacification efforts in general, making their application on a large scale nearly impossible.

The Khalq/Parcham split continued to pose a major dilemma for Soviet advisers and for Moscow. Amin had been a member of the Khalqi wing of the PDPA; Karmal, the leader of Parcham. The army and the forces of the interior ministry (Sarandoy) were primarily Khalqi, and its loyalty to Karmal was often in question. Moscow could not allow a purge of Khalqis, but also realized that Babrak Karmal and his Parcham faction were weak. KhAD was created, with KGB help, to replace the security agency that had functioned under Amin. The creation of a new agency had several purposes. First, it was meant to dissociate the security service from Amin's repressive rule. Second, it was to be a security service loyal to Karmal, not one in which Amin loyalists would undermine his rule. Mohammed Najibullah, a Parchamist, was installed as its head, and KGB advisers were sent to help him build up the agency.[77]

Although Parcham increasingly occupied the most senior positions, many lower-tier members, particularly in the army, were Khalqis.[78] This created additional friction within the party. Kim Tsagolov, a military adviser, told the deputy chief of the International Department in 1982 that installing Karmal was a mistake, "not because Karmal is not worthy of being a leader—he is a founder of the PDPA—but because there are many more Khalqis, and they are the ones spilling their blood, while many Parchamists are sitting in government offices, preferring to become apparatchiks."[79] Soviet advisers also began to split, some of them being more inclined to support Khalqis and others Parchamists. The KGB, on the whole, supported the latter, while the military supported the Khalqis, perhaps because they were the ones, as Tsagolov put it, "doing the fighting."[80] This split was noticeable in the early years of the war, but would become especially apparent even at the Politburo level when Najibullah took over and Soviet troops were withdrawing.

The divisions among Soviet representatives and advisers both mirrored and exacerbated divisions within the PDPA, the DRA government, and the armed forces. Of course, even if party unity had been achieved, it is far from certain that this would have enabled the party to make big gains with the population and attain the kind of legitimacy that would allow it to run the country peacefully. The party continued to exist primarily in cities; its presence in the countryside was largely on paper. At a meeting with Marshal Sokolov and Fikriat Tabeev, the Soviet ambassador, in March 1984, one party adviser admitted that in his region only 10 percent of the villages had any sort of PDPA presence. At the same time, he lamented, the center did not seem to mind that PDPA functionaries were not making their way into the countryside.[81] The figure of 10 percent was probably an estimate, and it covered only one region, not the country as a whole.[82] Nevertheless, it is clear that even by the fourth year of the occupation very little progress had been made in terms of "widening the social base of the party," a goal that Politburo leaders in Moscow had set in January 1980.

The difficulties in political work were similar to the ones that the Soviets faced in their effort to improve the military situation. The more Soviet advisers or troops became involved, which they were doing in order to stabilize the DRA government, the less the DRA government seemed able to act independently. The problem of restoring the Afghan army's ability to fight independently and of encouraging leaders to make decisions without turning to their Soviet tutors for help (a problem discussed later in this chapter, as well as in Chapter 3), was one of the major stumbling blocks in effecting a successful withdrawal.

The "nation building" described above was part of the Soviet Union's broader strategy in the first years of the war to stabilize the country. Moscow aimed to secure Karmal's position in the party while simultaneously building up the army and spreading the regime's influence. The limits of this approach were becoming evident early on, and by 1984 it was clear that the Soviet effort was stalemated on all fronts: the USSR was unable to reach an accommodation with the United States or Pakistan through diplomacy, to decisively beat the *mujahadeen*, or to work successfully with its Afghan clients to make their regime acceptable to the population.

Fighting a Limited War: The Military Dimension

It is almost axiomatic among senior Soviet officers who fought in the Afghan conflict, and then spoke or wrote about it, that the military was able to carry out its duty. Aleksandr Liakhovskii, who served in Afghanistan as part of the military advisory staff and later emerged as the most authoritative Russian writer on the topic, concludes that the Soviet military did not lose the war: "It would be wrong to say that the 40th Army sustained a military defeat. It is just that the army was faced with tasks which it was not in a position to carry out, since a regular army cannot radically solve the problem of revolt."[83] Lieutenant-General Boris Gromov, the last commander of the 40th Army in Afghanistan, goes further: "There is no basis for saying that the 40th Army suffered a defeat, just as there is no basis for saying we scored a military victory in Afghanistan." Nobody had ever asked the 40th Army to achieve a military victory, Gromov writes. Rather, the limited contingent was tasked with protecting the government of Afghanistan and preventing an invasion from outside, which it did.[84]

It is indeed true that the Soviet military never lost a battle or gave up a position in its war with the *mujahadeen*. It should also be noted that as woefully unprepared as the 40th Army was for the fight in Afghanistan in December 1979, it learned from its mistakes. Thus, for example, recognizing the deficiencies of its regular training program, the Soviet military set up specialized training courses for Afghanistan-bound soldiers in the Turkestan Military District, where conditions were similar to those found in Afghanistan. Similar training sites were set up in other parts of Central Asia. The army improved at fighting the *mujahadeen* as the war went on, and adjusted well to the introduction of new weapons such as the Stinger missile.[85]

Yet the decisive blow that the Soviet leadership hoped to strike at the opposition never came. Senior officers who served in Afghanistan were asked to carry out what they felt was an impossible task. Colonel General V. A. Merimskii, who served in Afghanistan in the early years of the war, writes that he repeatedly asked Marshal Sokolov, then the senior commander on the ground, to take this up with the defense minister, Dmitrii Ustinov. When Sokolov eventually did, Ustinov seemed to agree, but then asked Sokolov to at least find a way to close the borders and stop arms from entering Afghanistan: "All right, you can't deal with the counterrevolution, but can you defend against penetration from the outside?" Sokolov apparently replied that he could.[86]

Responses like Sokolov's probably helped to prolong the war, giving the Politburo reasons to believe that it would yet be possible to change the military situation for the better. In fact, the Soviet military was having trouble dealing with a task that could seem simple only to someone who had little idea of how weapons were crossing the border. Afghanistan's long mountainous border with Pakistan was almost impossible to control. Historically, Afghan kings secured alliances with Pashtun tribes living in the area and rarely sent their own regular army to guard the border with British India. Even then it was assumed that Pashtun tribes, who lived on both sides of the Durand Line, would be able to migrate back and forth as they saw fit. Ustinov's request to "close the border" sounded deceptively easy.

Moreover, Sokolov himself knew that even this "simplified" task was all but impossible to carry out. The Soviet military and the DRA tried to secure the border as best they could using a combination of military units, Afghan secret-police detachments, and border patrols. This still left gaps for penetration, on top of which the reliability of Afghan forces was always in question. In a meeting with Soviet advisers in March 1984, Sokolov admitted that, as far as closing off the borders was concerned, "at the current moment we cannot do it. Right now we have to close off the most important sectors."[87] Arms continued to flow into Afghanistan from Pakistan, completing a long supply chain that included US and Saudi funding and weapons, sometimes acquired from countries such as China, Egypt, and Israel.

The 40th Army tried to compensate for its inability to close the borders by interrupting the supply lines on the Afghan side of the border.[88] In the early years of the war, this involved a heavy reliance on fixed-wing aircraft to provide air support in raids on *mujahadeen* supply lines. Bombardment was supplemented by attacks from helicopter gunships and by means of mines, which were often dropped from the air along supply routes. These had the effect of wounding *mujahadeen* as well as crippling mules and camels that might be carrying supplies.[89] As with other cases when mines were used in warfare, they became a lasting hazard for civilians, yet another of the tragic legacies of the war. According to Major General Oleg Sarin and Colonel Lev Dvoretskii, some three million such mines were dropped or laid in the years 1980–1984 alone.[90]

Another major preoccupation was protecting Soviet-DRA lines of communication, which supported both the 40th Army and DRA forces and the cities. The only reliable overland route was a highway that ran from Termez to Kabul and connected the latter city with urban centers like Jalalabad and Herat, forming a horseshoe through Afghanistan.[91] Typically for guerrilla warfare, *mujahadeen* often attacked supply lines, which were particularly vulnerable on the difficult roads.[92] Even the Kabul-Termez highway was a major challenge to drivers, particularly in winter. One Soviet source described it as a road that winds "in steep and narrow hairpin turns, with a perpendicular cliff on one side and an abyss on the other."[93] Not only were Soviet soldiers vulnerable in such conditions; they often found it difficult to attack *mujahadeen*, who seemed to melt away into the mountains above.[94] At the same time, the vulnerability of the roads and their importance for Soviet military and economic aid meant that the number of troops that could actually be used for operations was often quite limited: it is estimated that some 35 percent of Soviet troops were being used to guard roads.[95] This, and the high rate of hospitalization due to disease among Soviet troops, limited the number of soldiers available for combat operations.[96]

Both fixed-wing aircraft and helicopters were of major importance in other kind of attacks on *mujahadeen* positions. Especially in the early years of the war, Soviet strategy relied on "hammer and anvil" operations involving massive attacks from the air and mechanized advances on the ground. The 40th Army also relied increasingly on helicopter gunships, which—compared to fixed-wing aircraft—could drop bombs from a lower altitude (and thus with greater precision), and could also strafe rebel fighters. Over time, too, the 40th Army increased its use of special forces (*voiska spetsalnogo naznachen'ia,* or SpetsNaz), which could be employed for targeted attacks against bands of fighters. On the whole, the Soviet military, which was geared toward conventional warfare with an eye on central Europe, adjusted well to the requirements of mobility that came with a guerilla war in a mountainous terrain.[97] Nevertheless, air power remained a key feature of Soviet combat tactics, as well as a way to supplement transport by road.[98]

In addition to the above, the 40th Army also took part in some 416 scale operations, such as Operation "Blow" *(Udar)* undertaken in November and December 1980. More than 16,000 troops advanced on *mujahadeen* positions in the "central" zone, killing an estimated 500 fighters and taking another 736 prisoner. A similarly large operation took place in the Kunar Valley, where Soviet and Afghan troops tried to clear out an area spread out over 170 kilometers; more than 11,000 highly trained *desant* troops (air-assault forces) were brought into the valley by helicopter. Perhaps most famously, the 40th Army undertook no less than *five* attempts to clear the Panjsher Valley of Ahmad Shah Massoud and his fighters. The problem with these operations was always the same, however: although the troops succeeded in clearing the area, they never had the numbers to hold it, and efforts to solidify political control never matched the *mujahadeen's* ability to reinfiltrate the area once the troops were gone. The repeated failures against Massoud only raised his stature, allowing him to extend his area of influence on the one hand and to assert more independence from Pakistan and the rest of the resistance on the other. "Throughout the whole of that war," recalled a captain who had participated in the first of the Panjsher operations, "practically every operation ended in the same way. Military operations began, soldiers and officers died, Afghan soldiers died, the *mujahadeen* and the peaceful population died; and when the operation was over, our forces would leave and everything would return to what it had been before."[99]

Coordinating these efforts with Afghan forces was always a challenge. The Afghan army had been decimated by the purges inflicted after the revolution and by the massive desertions that followed. Although the numbers recovered, few of its units were ever up to strength, and there was always a danger that some soldiers would desert in mid-battle. Moreover, many of the officers were loyal to one or another of the *mujahadeen* commanders and would inform them of planned offensives. One such officer was the head of Afghan military intelligence, who passed valuable information on planned operations to Ahmad Shah Massoud. What the Soviets found was that often the *mujahadeen* forces had largely dispersed before the operation had even started. Soviet officers would resort to deliberately misinforming the Afghans with whom they were going into battle, sometimes giving them the correct operational details only at the last minute. The result was mutual suspicion and recriminations when things went wrong.[100]

The larger dilemma that Soviet generals faced was how to use force against the insurgency without alienating the Afghan population. While Soviet forces were generally able to achieve objectives set out in specific operations, Soviet tactics often undermined the broader efforts to pacify Afghanistan. Attacks from the air, even well-targeted ones, inevitably hit civilians as well as fighters. Even with the transition from fixed-wing bombers to helicopter gunships, civilian casualties remained high. Soviet officers might say that the war had to be won politically, not militarily, but their tactics contributed to the political problem. When one considers the restraints imposed by the situation and by Soviet leaders themselves, however, there weren't many options available. A major decrease in the use of airpower would have meant a much larger invasion by ground troops, particularly in view of the fact that a great many were needed just to guard the supply routes and provide other support functions. More boots on the ground could have created its own political difficulties within Afghanistan, as well as within the USSR.

Moscow faced two related challenges in the war: minimizing its impact domestically and internationally, and at the same time preserving freedom of action. The military strategy was typical for counterinsurgency warfare: protect main routes, cities, air bases, and logistic sites; support the Afghan forces with superior air, artillery, intelligence, and logistic capabilities; and strengthen DRA forces so that they could fight without Soviet support.[101] The USSR avoided becoming overcommitted by limiting its presence in Afghanistan to roughly 120,000 troops, and it never expanded the war into neighboring Pakistan, thus avoiding some of the pitfalls of US strategy in Vietnam. It faced domestic pressures and international criticism, but not at a level that made an immediate change of course obligatory. Appreciating the limits and costs of the war is crucial to understanding why Gorbachev could afford to spend several years tinkering with the Afghan problem before bringing home the troops.

The intervention in Afghanistan complicated Moscow's foreign relations and created potential new challenges for domestic politics. Nevertheless, these pressures were well within the range of acceptable tension for Soviet leaders, for several reasons. First, the Brezhnev leadership in particular seemed willing to weather the diplomatic isolation and seemed to believe that it would eventually pass. (The grain embargo mustered by the Carter administration, for example, collapsed as Canada and Australia peeled away from it, and even US farmers pressured the Reagan administration to let them start selling to the USSR again.)[102] Second, militarily and economically, the war was costly but well within the means of the USSR's military-industrial complex.

While the material costs of the war were significant, they were little more than a dent when viewed in comparison to the overall Soviet military budget. Aid to Afghanistan constituted a significant, but not overwhelming, portion of the USSR's overall aid to the Third World at this time—estimated at $78 billion from 1982 to 1986. Taken together, aid to the DRA military and the expenses associated with Soviet military amounted to about 1.58 billion rubles in 1984, 2.62 billion rubles in 1985, 3.20 billion rubles in 1986, and 4.12 billion rubles in 1987, or roughly $7.5 billion over the four years.[103] By comparison, the entire Soviet military budget as late as 1989 was $128 billion.[104] Similarly, according to Russian government records, Afghanistan's debt to the USSR by October 1991 was 4.7 billion rubles—roughly half of India's, and about a tenth of the total debt owed to the USSR by developing countries.[105]

Furthermore, although the war certainly required an exertion of military power and a consequent loss of life at levels higher than any since the Second World War, it was far from unmanageable. The official tally, presented to the Central Committee the day after the last soldier left Afghanistan, counted 13,826 dead, 1,977 of those being officers. The 40th Army had also suffered 49,985 wounded, of whom 7,281 were unable to return to duty.[106] These were significant numbers, but not overwhelming for a large force like the Red Army. As at least one scholar has pointed out, Soviet casualties were comparable to peacetime losses due to accidents.[107]

Special mention needs to be made regarding the Stinger missiles and their supposed effect on the Soviet decision to withdraw from Afghanistan. Early in the war, the Soviet army came to rely quite heavily on close air support—that is, helicopter gunships, such as the Mil Mi-24 helicopters. The *mujahadeen*, fighting with small arms, had almost no way to counter this strategy. It was in response to this situation that the CIA, after offering several inadequate surface-to-air missiles (SAMs) agreed to provide the Stinger, a powerful heat-seeking antiaircraft weapon that was mobile enough to suit the conditions of guerrilla warfare.[108]

Yet while the Stinger, first introduced in combat in September 1986, did give the *mujahadeen* an important antiaircraft tool, it hardly changed the course of the war. It is true that Soviet pilots now had to fly at higher altitudes and occasionally abandon their missions, while civilian and military visitors to Kabul from that point onward observed with horror the "screwdriver" descent to the Baghram airbase that helped planes to evade the new weapon. Yet it is also true that the Soviet military and pilots adjusted, fitting aircraft with various devices to disorient the missiles, flying at night, or staying so low to the ground as to make the missiles useless. Although this adjustment allowed the Soviets to limit damage caused by the Stingers, it meant sacrificing accuracy and precision, and relying on even more damaging higher-altitude bombing.[109]

Finally, even though Afghanistan was on the Soviet border, there was little fear that the conflict would escalate to the point that the Soviet Union itself would be threatened. This was not for lack of trying on the part of the *mujahadeen*. From 1985 onward, in particular, the Pakistani ISI (intelligence service) formulated plans to attack targets within the Soviet Union. For example, in 1986 the ISI trained fifteen Afghan resistance commanders to launch attacks within Soviet territory that were aimed at disrupting the Soviet supply chain. While several attacks on the rail link between Samarkand and Termez (the last outpost on the Soviet border) were successful, others failed. In December 1986, there were also attacks on a power station in Tajikistan.[110]

On the whole, however, such successes were very limited precisely because of US and Pakistani fears that the war could potentially escalate into a wider conflict as a result. The USSR responded with force to these incursions, bombarding the Afghan side of the border heavily. When the

ISI commander in charge of aid to the resistance devised a plan to hit the "Friendship Bridge," which provided a road link between Afghanistan and the USSR, it was called off by Pakistan's president, Zia ul-Haq, who feared escalation of the conflict.[111] An April 1987 attack which destroyed several buildings on Soviet territory led to a Soviet protest which apparently caused some panic in the Pakistani Foreign Office. According to Mohammad Yousaf, the Soviet ambassador to Islamabad relayed the message that "if any further operation was conducted in the Soviet Union, the consequences for the security and integrity of Pakistan would be dire."[112] This prompted the local CIA official to ask Yousaf "not to start World War III" by conducting operations in Soviet territory. Eventually the attacks were called off completely.[113]

Although the war was costly in terms of lives and matériel, it was well within the limits of what the Soviet Union could manage.[114] In every sense, the Soviet intervention in Afghanistan was a limited war, much more so than the US involvement in Vietnam. The Soviet Union never expanded the war outside Afghanistan, even though it may have helped to destroy *mujahadeen* camps and intercept arms convoys before they crossed the border. Nor did Soviet leaders increase the number of fighting troops above a limit reached early in the war, even though it was obvious that there were often not enough troops to carry out the "hold" part of their clear-and-hold strategy. Soviet leaders maintained these limits because they saw their role as helping the Kabul government to establish its own defensive capability, rather than fighting the war on its behalf; because taking the war to Pakistan would have undermined their goal of gaining international legitimacy for their client and would have exacerbated the tensions caused by the intervention itself; and because they needed to keep the war limited enough to keep it semi-hidden from the Soviet public.[115] Keeping the war limited allowed the Soviet leadership to maintain greater freedom of maneuver in its decision making, as well as insist that the intervention was a "private affair" between two friendly states. Along with the (relatively) limited impact of the war within the USSR itself, restricting the scope of the war allowed Soviet leaders to delay the discussion about withdrawal.

Social Effects of the War

After the United States became heavily involved in Vietnam, political elites in Washington came under pressure from a highly motivated antiwar movement and from widespread disillusionment with the war. The war brought about the downfall of an otherwise popular president, Lyndon B. Johnson. The effects of the Afghan war have been harder to gauge. Did it serve as the genesis of problems within the military, such as drug use and hazing *(dedovshchina)*, or did it worsen problems already there? Did it alienate Soviet Muslims, or did that alienation come later, and for other reasons? Did the war serve to break the faith of the Soviet citizenry in its main institutions, especially the army, or did this happen later, for reasons only tangentially connected to the war? The difficulty of answering these questions is compounded by the fact that the war straddles the pre-perestroika period, the period of reforms, and the beginning of the USSR's disintegration. It is tempting to look to the war as the source of many of the troubles that led to the collapse of the Soviet Union. But one must be careful not to inflate the war's significance, even in the tumultuous period of 1989–1991. Most crucially, however, one must understand that both the social effects of the war and opposition to the war among the Soviet population (to the extent that it existed) were scarcely felt by the USSR's leaders, and thus played little if any role in their decision making on the war.

During the course of the conflict, the broader Soviet public, through unofficial channels, learned about the war and began to voice its disapproval. Soviet leaders were not unaware of this widespread frustration. In July 1981 the Politburo was already considering how to handle the letters coming to the Central Committee from parents and relatives of the fallen.[116] As early as 1983, Yurii Andropov was worrying about the effect of the war on Soviet society. Yet it is one thing to say that the effect of the war on public opinion *concerned* Soviet leaders, and quite another to say that it inspired them with a sense of urgency about withdrawal. The nature of the pre-glasnost Soviet system meant that it was possible to conceal the details of a "limited war" from broader Soviet society, at least for a few years. When the Politburo considered "perpetuating the memory of soldiers who died in Afghanistan," its members seemed to have few qualms about keeping any potentially sensitive information off the soldiers' gravestones, since "from a political point of view this would not be entirely good."[117]

Indeed, in the early years of the war, its presence was clearly felt only in a few areas. In Tashkent, often the first stop for returning veterans as well as for troops heading to Afghanistan, the sight of wounded young men was familiar. Svetlana Alexievich, the well-known Russian human-rights activist and author, described the city airport in 1986 as a place where "young soldiers, no more than boys, hop about on crutches amidst the suntanned holiday crowds." She goes on to say that nobody noticed the soldiers; they were "a familiar sight here, apparently."[118] Most cities were not so closely linked to the war, however.

As more and more soldiers completed tours of duty and returned home —sometimes as wounded veterans, sometimes as bodies for burial, but almost always marked by the experience—the war became increasingly difficult to keep a secret.[119] Initially it was mostly discussed on the pages of dissident *samizdat* publications, such as the *Chronicle of Current Events (Khronika Tekuschikh Sobytii)*. Antiwar posters and leaflets were noted in 1981 in several major cities, including Moscow and Leningrad.[120] By 1985, opposition to the presence of Soviet troops in Afghanistan was on the rise among a wider sector of the public. Anatolii Cherniaev noted early in April of that year, in his diaries, that a "torrent of letters" about Afghanistan was coming in to the Central Committee and the editorial offices of the daily newspaper *Pravda*. Unlike earlier letters, which had often been anonymous, these were signed. Letters came not only from relatives of servicemen but from members of the wider public. Soldiers and even senior officers were writing as well, and one general wrote that he could not explain to his subordinates why they were in Afghanistan.[121] As we shall see, however, it was Gorbachev's decision to open up reporting on the war that truly brought it into the public domain—and even then, public opinion does not appear to have played a role in Soviet decision making.

Another potential area of concern was how Soviet Muslims, particularly those in Central Asia, would respond to the war. Many were called upon to participate in the war, either as soldiers or as translators and advisers. Others came into contact with the war and developments in Afghanistan through the influx of students and trainees that came to Central Asia, particularly Tashkent. Indeed, after the invasion of Afghanistan, some Western scholars speculated that Soviet Muslims might rise up against the state.[122] Yet the idea of a Soviet Muslim revolt sparked by the Soviet invasion was largely a fantasy of the CIA and some sympathetic scholars in the West; in practice, the Afghan war did not greatly change the religious climate in the Central Asian republics, and the two issues never seem to have become associated in the minds of the Soviet leadership. Still, it is worth pausing to consider what links the war had with later developments in Central Asia, and why the war did not cause the kind of "Muslim revolt" that US hardliners hoped for.

There were several reasons to suspect that the war would cause unrest among the Soviet Union's large Muslim population. The first is that the Soviet house always stood on a somewhat shaky foundation when it came to maintaining harmony among the nationalities and keeping those populations loyal to the state. The second is that eradicating Islam and bringing it within the control of the state had always been more difficult than doing the same with Orthodox Christianity. The third is that three Central Asian Soviet republics each shared a major border with Afghanistan: the Uzbek SSR, the Tajik SSR, and the Turkmen SSR. In all three cases, the dominant ethnic groups of the republic lived on both sides of the border and could migrate with relative ease. Yet resistance to the war was almost as minimal in Central Asia as it was elsewhere in the USSR. A brief review of the region's history under Soviet rule is useful for understanding why that was the case.

By the late 1970s, the Central Asian states were, on the whole, well integrated into the USSR. The peoples of the Central Asian republics had been incorporated into the Soviet state in the 1920s through a mixture of co-optation and often brutal counterinsurgency.[123] The peoples of the region could point to the benefits (widespread literacy, access to medicine, irrigation) and costs (overreliance on a "monocrop," namely cotton; environmental degradation; bureaucratic corruption) which came with Soviet modernity. Yet in the postwar era, a certain equilibrium had been established. With regard to religion, the state returned to the policy of limited control and benign neglect, setting up institutions to monitor and supervise religious activity. Soviet Muslims responded to "official" religion in different ways. Many turned to "parallel" Islam—praying in unregistered mosques set up in apartments or abandoned buildings, or joining unsanctioned religious groups, such as Sufi circles. Religious groups also fulfilled certain functions, like conducting wedding and funeral ceremonies, which were normally carried out by government offices. At the same time, many local communists in Central Asia and the Caucasus also participated in "parallel" Islam. The religious groups rarely engaged in political activity per se. Their challenge to the Soviet state was limited to their existence outside bureaucratic and ideological control.

While primarily quietist, "parallel" Islam could provide a forum for more overt anti-Soviet activity. Examples of this were the *samizdat* publi-

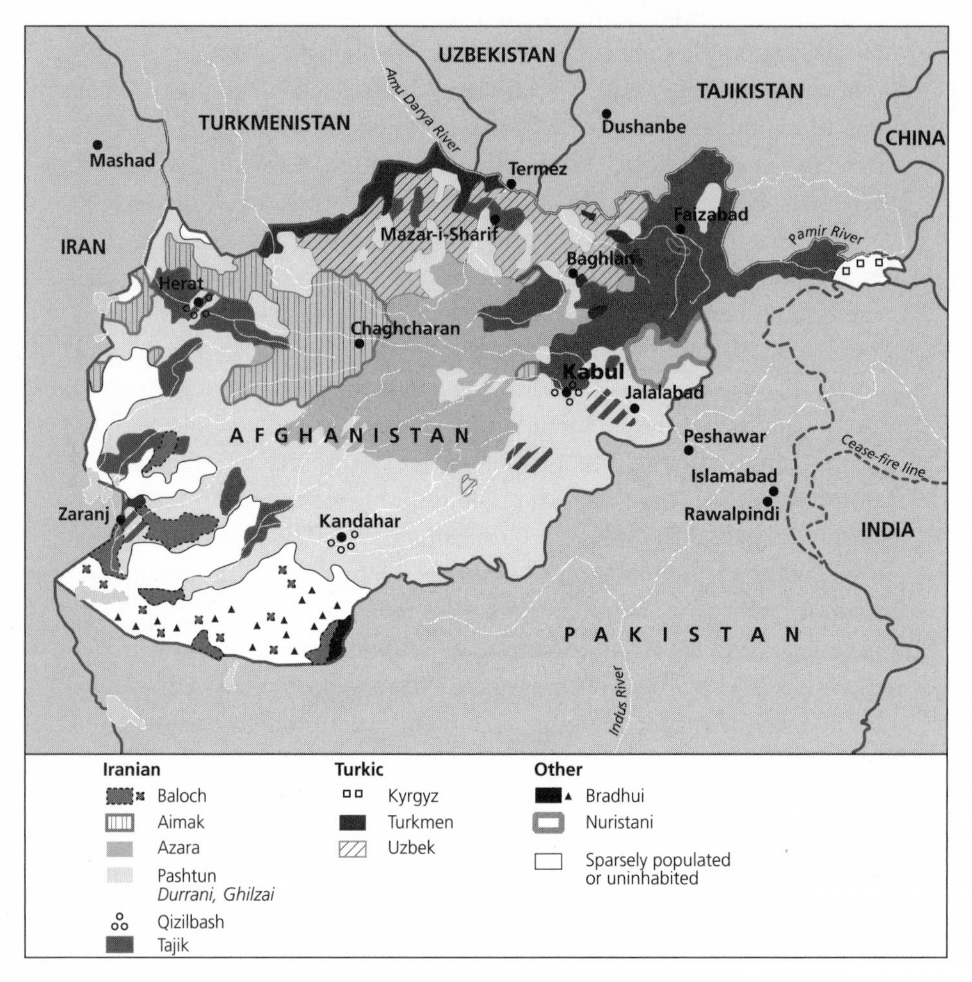

The ethnic groups of Afghanistan

cations which circulated in both Central Asia and the Caucasus and often carried truly anti-Soviet messages, including calls to avoid military service.[124] The implications of these messages were highlighted during the Afghan war. On several occasions, there were reports of violence breaking out between Muslim recruits and military authorities. In June 1985, there were reports of clashes between Chechen recruits and officers when the former refused to go to Afghanistan and fight their "Muslim co-religionists."[125] This was not an isolated incident. Later that summer, a military train carrying conscripts from the North Caucasus to the Afghan border was delayed when a fight broke out between the Muslims and the Christians on the train, with arguments about religion fueling the entire incident.[126]

Cross-border ethnic and religious ties on the one hand and the existence of a "parallel" religious network on the other gave Soviet Muslims, particularly in Central Asia, the ability to discuss the war more openly than could others in the Soviet Union. Not surprisingly, some of the earliest strong negative reactions to the war came from among Central Asian Muslims. In 1983, for example, the Council for Religious Affairs (CRA) reported to the CPSU Central Committee that in Tajikistan, some unregistered mullahs were issuing statements saying "it is forbidden to bury Soviet soldiers killed in Afghanistan according to Muslim rites, because they fought against true Muslims."[127]

The war permeated Central-Asian Muslim communities in other ways as well. Some of the resistance organizations in Afghanistan were able to find supporters north of the Soviet-Afghan border. In 1983 the Tajik-dominated Jemiat-e Islami claimed to have 2,500 members in Soviet Tajikistan.[128] The number may have been an exaggeration, but any presence at all is significant and would have been of concern to Soviet authorities. Later in the decade, thousands of Central-Asian Muslims would travel, clandestinely, to madrasas of the Deobandi school in North-West Pakistan, where they were supported and even given scholarships.[129]

The potential to stir trouble among the Central-Asian Muslims proved tempting for the CIA and the Pakistani ISI. In 1982, pamphlets with titles like "The Life of the Great Muhammad" and "How to Pray," as well as books such as *Islam and Social Justice*, by the Pakistani Islamist Sayed Abul-ala al Mawdoodi, were being printed in Peshawar in Russian and smuggled into Central Asia. Their existence came to light when they were criticized in a Kyrgyz newspaper by a local academic.[130] The CIA decided to back these efforts as part of its support for the anticommunist resistance. As CIA chief William Casey saw it, the Muslims of Central Asia "could do a lot of damage to the Soviet Union."[131] The intelligence agencies experimented with a small-scale infiltration, coordinated by Mohammad Yousaf. With CIA help, some 10,000 copies of the Koran were prepared in Uzbek, along with books describing Soviet atrocities against Uzbeks. During the summer of 1984, dozens of *mujahadeen*, primarily ethnic Uzbeks, made the night journey across the Amu Darya (Amu River) to bring the books into Uzbekistan. According to Yousaf, the Koran was well received, but there was little interest in the books on atrocities.[132] A March 1984 CRA report noted that growing ties with Afghanistan had led to an increase in religious literature coming into the Tajik SSR.[133]

Soviet officials at all levels were certainly aware that the Afghan war could make the situation in Central Asia and other Muslim republics more difficult. The possibility of the "Islamic factor" being used by "enemies" to destabilize the Soviet Union was the subject of a 1981 Central Committee resolution, and the CRA was tasked with helping to neutralize the threat.[134] Officials in Uzbekistan also noted with alarm that "agents of imperialism" were trying to use the "Islamic factor" in the republic, not only stirring up religious feeling, but giving it an "anti-Soviet, nationalist direction."[135]

As the war dragged on and the number of Soviet troops who had served in Afghanistan increased, public awareness of the war increased as well—but this did not translate into widespread opposition to it. CIA compilations of disturbances associated with the Afghan war for the period 1984–1987 show that they were spread throughout the Soviet Union. The non-Russian republics seem to be more heavily represented, but the disturbances were by no means confined to predominantly Muslim areas.[136] The CIA assessment also conceded that in Central Asia support for the war had "increased markedly, while opposition has only grown marginally." The CIA's sources were apparently repelled by the violence of Islamic fundamentalists and feared the consequences if they were to seize control in Afghanistan.[137]

Indeed, while a thorough study of Soviet Muslims' attitudes toward the war has so far been impossible, there is evidence—admittedly inconclusive—that rather than viewing the Soviet intervention as an attack on fellow Muslims or on people of the same ethnicity, young people in Central Asia generally accepted official Soviet explanations for the military involvement. In the war's early years, some even volunteered or thought of volunteering, identifying the "internationalist" action there with the legends of heroism they knew through popular songs such as "Grenada," which celebrated Soviet volunteers in the Spanish Civil War. On the less romantic side of things, Afghanistan's seeming availability of consumer goods, brought back by returning soldiers, also increased the appeal of serving there.[138] What is particularly interesting is that this attitude was shared not just by the children of Soviet elites, but even by those who lived in less privileged sectors, participating in clandestine religious groups—the very people who would become the nucleus of a nationalist or religious opposition later in the decade. Mehdid Kabiri—who was a member of a clandestine religious study circle, later a political activist, and eventually a leader of the Islamic Renaissance Party of Tajikistan—even volunteered to serve in Afghanistan while performing his military service, though he was not ultimately sent.[139]

Among the soldiers who went to fight in Afghanistan, there was never any large-scale defection of Muslims to the *mujahadeen.* True, some did go over to the other side, usually after being held in POW camps—but so did a number of Russians. The total number of such defectors was several hundred at most. Viewed against the hundreds of thousands of troops who went through the war over ten years, the number is not significant. Nor does there seem to have been a particular problem of draft avoidance by Central Asians as a result of the war; while precise figures are unavailable, Mark Galeotti has shown that participation in the war was consistent with the makeup of the Soviet armed forces generally. The Baltic republics were heavily underrepresented, as were the Transcaucasian ones; but the Tajiks and Uzbeks participated in numbers proportional to the size of their countries' populations.[140] They were less likely to serve in combat roles, but this is primarily because they were underrepresented in elite units, which were dominated by Slavs.

Thus, Moscow did not respond to reports of the war's echoes among Soviet Muslims with any great alarm. At most, officials in Moscow felt some concern, and passed resolutions in favor of better propaganda and educational efforts. None of the available Politburo records suggest that this was a primary concern for leaders in the years 1980–1986.[141] In light of what has happened in Central Asia since 1991, particularly in Tajikistan, it may seem strange that Moscow did not respond with greater alarm to the possibility of "blowback." There are several important factors to consider, however. First, the problem of religiosity among Soviet Muslims predated the Afghan war and had reached an equilibrium that seemed to satisfy both the state and the religious community. The dissemination of literature from Afghanistan did not affect it greatly.[142] Second, while Soviet Muslims were probably not supportive of the war, this did not mean they were ready to launch a holy war against the Soviet state or indeed to upset the social order from which they in some ways benefited. Finally, to the extent that officials were concerned about the "Islamic factor" being a destabilizing consequence of the war in Afghanistan, they believed that the best way to prevent it was to win the war and establish a stable government in Kabul.[143]

The difficulty of explaining the war to the public, and concerns about the broader social damage it could cause, thus clearly weighed on the minds of Soviet leaders from the early years of the conflict. But there was no mass organized opposition to the war, nor did one loom on the horizon. Soviet leaders could conduct their Afghan policy in the confidence that for the vast majority of citizens, attitudes toward the war were at worst ambivalent.

During the first months of 1980, the key Soviet foreign-policy decision makers came to a consensus that Soviet troops would have to play an active role in Afghanistan and remain there indefinitely. Having taken the momentous step of intervening in the country, Soviet leaders also raised the stakes. At this point, a withdrawal before Moscow's goals were achieved, or one followed by a collapse of Karmal's government, might be seen as a defeat of the Soviet military. Such a defeat would be a blow to Moscow's prestige in the Third World and to its sense of parity with the United States, both of which Soviet leaders valued highly.

Between the invasion in January 1980 and the death of Konstantin Chernenko in 1985, Moscow pursued a three-track policy to stabilize the Karmal regime and normalize the situation in Afghanistan. The first track was military. Soviet leaders accepted the need for the Soviet army to engage the Afghan opposition directly, so as to protect their client government in Kabul. They accepted that it would be necessary to do this until Afghan security forces could fight on their own. The second track was the effort to unify the PDPA and make it more acceptable to the population. The third track was diplomacy, which in this period meant participation in the United Nations' effort, discussed in the following chapter. Diplomacy could secure greater recognition for the regime, as well as stop interference from Pakistan.

The Soviet "nation-building" campaign reflected the confidence of Soviet leaders that their experience of exporting modernity could help them win in Afghanistan. Soviet advisers had been active in Asia, the Middle East, Africa, and Latin America since the 1950s, and particularly during the 1970s. Soviet economic aid and expertise helped postcolonial governments to gain and maintain legitimacy. It is not surprising that Soviet leaders tried to draw on this experience in Afghanistan. Thus, while political advisers made sure the PDPA followed a moderate path and helped Afghan activists to spread the government's influence in the countryside, technical specialists and economic advisers tried to bring some of the benefits of modernity to the Afghan people. These efforts did bring some real benefits to many Afghans, but the overall strategy failed to win their allegiance to the Kabul government.

2

THE TURN TOWARD DIPLOMACY

In November 1982, after eighteen years in power, Leonid Brezhnev died. His passing was hardly unexpected: for years, the General Secretary had been in evident physical and mental decline. His condition had served as great fodder for joke tellers, but in reality watching him stutter and stumble through his public appearances—his eyes sad, frightened, or simply vacant, his body staying upright only with the support of bodyguards—was depressing. Because his leadership was associated with the decision to intervene in Afghanistan, his death brought hope that perhaps a new approach could be found to the Afghan problem.

Brezhnev was succeeded by Yurii Andropov, one of the most enigmatic figures in the Soviet leadership. Many saw him as a hardliner on both domestic and foreign issues (and among the chief instigators of Soviet interventions, not just in Afghanistan but also in Hungary and Czechoslovakia). Others noted his asceticism, his love of Western culture, and his moderate approach to international relations. In any case, during his brief hold on power, from 1982 to 1984, he made attempts to battle corruption and breathe new life into the Soviet economy, and strove to reduce tensions abroad. This was true for Afghan policy as well: Andropov, who played a key role in the decision to intervene, also led the effort to find a way out through diplomacy. But if Andropov's mind was vigorous, his physical health was not. About a year after he took office he became seriously ill, and for the last months of his life he ran the country from his hospital bed. When he died in February 1984, he was succeeded by Konstantin Chernenko, a servile apparatchik who had grown close to Brezhnev toward the end of the latter's life. Chernenko was, if anything, in even worse shape than Brezhnev had been in his final years. The penultimate General Secretary was never really at the helm, being far too ill to attend Politburo meetings regularly and not a forceful enough personality (or a strong enough intellect) to guide affairs from his sickbed, as Andropov had done.

The Soviet Union's turn toward diplomacy was not just a result of the power transition after Brezhnev's death. Rather, it came from an emerging consensus that the war could not be won. At first, Moscow had assumed a defensive attitude in the face of the nearly universal condemnation of its intervention, and had undertaken a number of propaganda and diplomatic efforts to counteract the hostility of the United States, Western Europe, and much of the Muslim world. Within the USSR itself, there was no public protest, but there was a strong reaction to the invasion by some of the more well-connected intellectuals in Moscow, as well as party members and senior military officers. Initially ignored, their criticisms eventually came to be heard by the USSR's key foreign policy makers. By 1982, Soviet leaders had come to accept the need for UN diplomacy to help resolve the Afghan conflict, and prodded their Afghan clients to do the same. But the high level of US-Soviet tensions in 1983 scuttled these efforts, and confusion at the top of the Soviet hierarchy caused by Andropov's death and Chernenko's illness meant that no further significant initiatives were taken before 1985.

The Soviet Intervention and the Carter Doctrine

The Soviet intervention in Afghanistan brought cries of protest from Western capitals as well as the Muslim world.[1] It damaged Moscow's relationship with friendly nations such as India, which were troubled by the precedent of an invasion aimed at regime change.[2] At the UN General Assembly, the Nigerian representative "expressed disappointment at the Soviet action, since . . . no country had assisted the Third World more."[3] A resolution condemning the Soviet intervention was adopted by a vote of 104 to 18, with 18 abstentions. The invasion also contributed to frictions with China, whose leaders made the withdrawal of Soviet troops one of the preconditions for an improvement in relations.[4]

The strongest reaction came from the United States. The Carter administration—in particular, Zbigniew Brzezinski—was determined that the Soviet move could not go unpunished. Following several weeks of internal debate, the administration agreed on a set of measures, including a tight ban on the export of high technology, the suspension of US-Soviet official exchanges, a limit on grain exports to the Soviet Union, the boycotting of the Moscow Olympics, and the withdrawal of SALT II from consideration in Congress. On January 23, in his State of the Union speech, the president laid out what became known as the Carter Doctrine. "Let our position be absolutely clear: An attempt by any outside force to gain control of the Persian Gulf region will be regarded as an assault on the vital interests of the United States of America, and such an assault will be repelled by any means necessary, including military force."[5]

Individually, the effect of these measures on Moscow was minimal. The international grain boycott that Washington tried to organize soon eroded, since the trade was profitable for farmers, and SALT II had already failed in Congress the previous summer. As Jack Matlock, a career diplomat and one of the most knowledgeable US experts on the Soviet Union, wrote, "All of these moves were damaging to Soviet prestige, but they were not sufficient to convince the Soviet leaders that they had anything to gain from withdrawing from Afghanistan before they had accomplished their purpose."[6] More significant in the long term was that the invasion brought the US and Pakistan closer together after a low point in their relations, and that the administration became committed to supporting the Afghan resistance.

Still, the international reaction to the invasion must have felt like a tidal wave to Soviet diplomats in the first weeks of 1980. The United States could not be allowed to use this moment to turn Moscow's allies against the USSR. In March and April 1980, the Politburo approved a counterattack to "increase the activity of the international public against the aggressive activities of the USA in the Persian Gulf."[7] Aside from the mass-media offensive, the Politburo approved in March a plan of measures "for the activation of the international community against the aggressive actions of the USA in the Persian Gulf area."[8] The Soviet plan envisioned the activation of every party organ that could possibly be of relevance, including the International Department, the Komsomol, and "committees of solidarity" with individual Asian and African countries, as well as with Palestine. Soviet leaders would reach out to the nonaligned movement at its twenty-five-year anniversary conference, and would use various international peace and trade union conferences to organize resolutions against US policy in the region.[9] Soviet Muslim clergy issued statements addressed to the Muslims of the world, asking them to remember that the USSR had always been a friend of Muslim peoples and their defender against imperialism.[10] But the limitations of the strategy were evident at the "Tashkent Conference," assembled in September 1980. Although the occasion of the conference was supposed to be a celebration of the 1,500-year anniversary of the Hejira (emigration of the prophet Muhammad to Medina), the real purpose was to push "anti-imperialist" propaganda in the wake of the Soviet invasion. A planning meeting of four Soviet muftis in January 1980 issued a declaration against "US imperialists, Israeli Zionists, the traitor [Anwar] Sadat, and Chinese hegemonists' meddling in Afghan affairs."[11] Only a few of the seventy-five countries invited sent delegates, and some of those that did were vocal in their criticism of the Soviet invasion.[12] In general, efforts to rally Third World states to Moscow's side failed to counter the campaign spearheaded by the United States and Pakistan. In 1981, a resolution condemning the invasion was once again approved in the UN General Assembly by an overwhelming vote, a pattern that would hold for the remainder of the war. The Conference of Foreign Ministers for the Nonaligned States and the Organization of the Islamic Conference both called for a withdrawal of foreign troops from Afghan soil.[13]

At the same time, the Politburo directed the KGB to carry out appropriate measures along a similar line in developing countries, especially Iran.[14] It is not clear what exactly this meant, and there are no KGB documents that provide any detail. It is almost certain, however, that the KGB's task involved more clandestine ways of achieving the same goals the diplomatic initiative was supposed to help deliver. According to Leonid Shebarshin, the KGB resident in Tehran from 1979 to 1983 and later chief of the First Directorate, the KGB's instructions (aside from the order to gather information) were to try to "increase anti-American feeling and soften anti-Soviet feeling." The latter was nearly impossible: the invasion of Afghanistan had helped to make the USSR almost as big of an enemy as the United States, in the eyes of revolutionary Iran.[15] Nevertheless, Soviet leaders remained hopeful that the anti-Americanism of the revolutionary government would help to neutralize the Carter administration's efforts in the Persian Gulf.

In the months after the initial intervention, Soviet leaders still hoped they could enable the Afghan government to regain control of the country. As such, they rejected the proposals that came from a number of European countries for a political settlement in Afghanistan.[16] These plans called for a political resolution within the country, a withdrawal of Soviet troops, and mutual pledges of noninterference. The USSR rejected these approaches because they emphasized a political settlement within Afghanistan that would be arranged by outside powers; they implicitly questioned the legitimacy of the Karmal government; and they threatened to limit Moscow's freedom of action. The Soviet Union was interested only in proposals that would lead to a pledge of noninterference by outside powers. Such proposals would also have implicitly recognized the legitimacy of the Karmal government.[17] To discuss the Afghan situation without including the Afghan government—as a Soviet diplomat told his French colleague—would be "absurd." There could not be any settlement of the Afghan problem if the "real existing power in Afghanistan is ignored."[18]

The Soviet leadership was not completely rejecting the possibility of using diplomacy to settle the Afghan question. Rather, Soviet leaders wanted to avoid any approach that might undermine the legitimacy of the Kabul government or the USSR's actions in support of it. They hoped to use the US offensive in the region to deflect some of the anger from Muslim countries that was directed at the USSR following the invasion. As the Afghanistan Commission put it, the question of US bases and troops in the Persian Gulf region needed to be raised repeatedly. This would allow the Politburo "to widen the circle of countries well disposed to our position in Afghanistan, or, at least, inclined to approach it with understanding."[19] In other words, the Soviet leadership aimed to use the US offensive in the region to deflect some of the anger from Muslim countries directed at the USSR following the invasion.

Growing Opposition to the Intervention

Georgii Kornienko, a Soviet deputy foreign minister who later became directly involved in Afghan affairs, wrote that at the June 1980 plenum no one had spoken out against the invasion or even raised a question about it. Perhaps if they had, he suggested, the Politburo would have started looking for a way out earlier.[20] In fact, although there had been no opposition at the plenum, by June 1980 a number of party and state officials, as well as leading figures of the academic world, had made their concerns known to Brezhnev and others in the leadership.[21] There was also dissatisfaction in the military, not just among those who had opposed the invasion in the first place, but also among those who had gone to Afghanistan and taken part in the fighting. Over the next several years their calls for a change of strategy began to accumulate, and by early 1981 even the defense minister, Ustinov, a supporter of the intervention in 1979, was willing to approach other members of the leadership with the idea of withdrawing troops.

One of the earliest and most eloquent known criticisms of the intervention from within the Soviet elite came as early as January 20, 1981, when the Institute of Economics of the World Socialist System sent a memorandum to the Central Committee of the CPSU, as well as to the KGB. The memorandum, signed by Oleg Bogomolov, director of the institute, argued that the invasion had done a great deal of damage to Soviet interests and to détente, as well as giving new stimulus to the opposition, which was now able to call the population to rise up "against a foreign invader." The regime was isolated, able to count on support only from the socialist camp, primarily from the USSR. The memorandum listed eleven ways in which the invasion had damaged Soviet interests, including the effect it would have on the arms race, the economy, and Sino-Soviet relations. Coming at a time when the extent of resistance was perhaps not yet clear, it included a prophetic note: the leadership needed to maneuver for a way out prior to the start of spring, when warmer weather would bring increased attacks and Soviet troops would be drawn into the fighting.[22]

Anatolii Cherniaev's diary entries for the winter and spring of 1980 record the disgust and worry among his circle of "party intellectuals," historians, and others with academic training working in the International Department and elsewhere in the Soviet apparatus.[23] Although many of these people probably never made their views known outside a small circle of friends, some appealed to the Central Committee and even to the General Secretary himself.[24] Those who traveled abroad experienced first-hand the strength of the international reaction. Georgii Arbatov and *Pravda* correspondent Yurii Zhukov, returning from a trip to Italy, where they had met with American academics, secured a meeting with Brezhnev in May 1980 at which they tried to convince him of the damage the invasion had done to US-Soviet relations.[25] Similar efforts were undertaken by specialists on the region.[26] Yet the effect of these early petitions was clearly minimal. The views of even the most respected academics could not compete with the views of the party's most senior leaders.[27]

Troubling information also came from Soviet journalists who were sent to Afghanistan to report on the progress of the revolution. Although they were limited in terms of what they could actually publish, some of them sent more truthful accounts using confidential channels. One correspondent wrote a scathing assessment, addressed to the CC CPSU, almost two years into the Soviet occupation, saying that military operations were largely counterproductive. Although the PDPA still controlled only 15 percent of the country, the operations against rebels only aggravated the relationship with the peasantry: "The tactic of pursuing the rebels and destroying their nests on their home ground is facing increased resistance from the local population. In the course of those operations, houses and crops are often destroyed, civilians are killed, and ultimately everything remains the same. The rebels come back and reestablish control over the territory."[28] The letter also contained some thinly disguised criticism of the attempts to paint the war as a battle against outside aggression, pointing out that this was in fact a "civil war" before anything else.[29] With its stark description of the Soviet army fighting against the civilian Afghan population, the letter echoed the nightmare scenario discussed at Politburo meetings in March 1979, when Soviet leaders had decided *not* to send in troops.

The most difficult to ignore were the concerns of senior officials in the Ministry of Foreign Affairs or the military. Mikhail Kapitsa, a long-serving diplomat and deputy minister at the time of the invasion, pointed out at a Foreign Ministry collegium meeting that the Soviet intervention would face enormous difficulties, and cited the experience of British troops in the nineteenth century. Gromyko reportedly asked, "Do you mean to compare our internationalist troops with imperialist troops?" Kapitsa replied: "No, our troops are different—but the mountains are the same!"[30] A number of other senior officials also expressed their concern, either to the minister personally or in written form.[31]

On the military side, senior commanders had expressed their opposition even prior to the invasion. As early as 1980, Marshal Ogarkov, General Varennikov, and General Sergei Akhromeev agreed to a certain extent that there was no military solution to the unfolding situation.[32] Yet negative assessments did not always make it all the way to the Politburo. General V. A. Merimksii, deputy chief of the Ministry of Defense operational group in Afghanistan during the early years of the war, writes that although Marshal Sokolov agreed with the assessments of field commanders who thought there was no military solution, the Politburo was unwilling to consider a pullout.[33] At times, however, senior officers were more optimistic about their prospects for defeating the insurgency. At a meeting at the Soviet Embassy in Kabul in January 1980, Sokolov said that the "counterrevolution" would be defeated by June first of that year.[34] Aleksandr Maiorov, the chief military adviser in 1980–1981, likewise said he believed the war could be won by the end of 1981.[35]

It is difficult, if not impossible, to evaluate which of these reports, if any, had an impact on the key decision makers: Andropov, Gromyko, Ustinov, Brezhnev, and to a lesser extent Ponomarev. It is not even clear which reports traveled up the bureaucratic chain. Leonid Shershnev, a lieutenant colonel and political officer posted to Afghanistan in 1981, sent numerous reports to his superiors (and, he later said, straight to Moscow) arguing that the Soviet army was doing more harm than good. There were no replies, and he was repeatedly warned not to bypass his superiors. Even Akhromeev, who seemed to agree with him, told him to steer clear of politics, which were not the army's business.[36] Since Akhromeev was one of the officers expressing his doubts to Sokolov, Ogarkov, and Ustinov, it is possible that Shershnev's concerns were made known, at least indirectly, even at the Politburo level. But it is also likely that many of these reports never made it all the way to the top decision makers—that they were intercepted along the way by "gatekeeper" subordinates who did not wish to anger their bosses with bad news.

Clearly, however, at least some of these views were filtering through to top Soviet leaders. By early 1981, doubts about continuing the intervention had started to form among Politburo members. Minister of defense Ustinov, who had rejected the officers' concerns prior to the invasion, now started to take them to heart. He was the one receiving assessments regularly from commanders in the field and knew first-hand the difficulties they were facing. In an interview with journalist David Gai, General Norat Ter-Grigoriants recalled a meeting with Ustinov early in 1981 when the latter asked: "In all honesty, when will we end the war there?" Ter-Grigoriants replied that it was impossible to "resolve the Afghan problem by military means" and recommended the formation of a coalition government.[37] In February of that year, Ustinov circulated a letter within the Politburo stating that "no military solution to the war was possible and that it was necessary to find a political and diplomatic way out." Yet no one else in the Politburo backed Ustinov, and the letter was never put on the agenda of a Politburo meeting.[38]

It is also clear that Brezhnev himself was troubled by the prospect of a long-term intervention and seriously concerned about the recent deterioration in East-West relations. Throughout the 1970s, he had been passionate about détente, even facing down Politburo colleagues when they opposed concessions he was willing to make in negotiations with the United States.[39] He hoped that Soviet troops could be brought home within a few months.[40] At a meeting in May 1980, Brezhnev listened to Valéry Giscard d'Estaing's criticisms of the Soviet invasion and to Gromyko's formulaic retorts, then asked to see the French president in private. When the two were alone, he said that he agreed with Giscard's views. While justifying the need to remove Amin, Brezhnev added that he knew Soviet troops could not stay in Afghanistan and that a political solution was necessary. Brezhnev continued in emotional tones: "I also wanted to tell you this, one on one. The world is not in universal agreement [with our actions]. I will make it my personal business to impose [a political] solution. You can count on me!"[41] Most likely, Brezhnev really did believe that Soviet troops should leave Afghanistan as quickly as possible. But he also listened to his main foreign-policy advisers, who did not believe that withdrawal was possible at this stage.

As time went on, new reasons emerged for Soviet leaders to consider withdrawal. It was becoming more difficult to keep the war a secret from Soviet citizens. Although the press still spoke only of limited Soviet aid and although there was a news blackout on the 40th Army's activities there, rumors had begun to spread. These were perpetuated by citizens who listened to foreign broadcasts. Also contributing to the rumors were the parents of soldiers who were wounded or had died in Afghanistan. By July 1981, Politburo leaders were worried about the consequences if information about the war became common knowledge, and were unsure how to handle the letters coming to the Central Committee from parents and relatives of the fallen.[42] Even the gravestones of fallen soldiers were prohibited from stating how or where they died. Mikhail Suslov, the chief Soviet ideologist, pointed out that any mention of the war on the headstones could have unwelcome consequences: "If we perpetuate the memory of soldiers who died in Afghanistan, what will we say on their epitaphs? In some cemeteries, there could be several such headstones—

so from the political point of view, this would not be entirely good." Andropov agreed.[43]

By the end of 1981, a significant shift had taken place in how the USSR's top foreign-policy decision makers thought about the war. The intervention of December 1979 had now lasted two years, and the concerns of their subordinates were becoming harder to ignore. The illusion that the invasion could help Moscow achieve its goals within a reasonable time frame faded. From January 1980 to the end of 1981, there were plenty of indications that the war was going poorly and that the intervention had not been worth the strain it had put on the Soviet Union's relationship with the United States, other Western countries, and the Muslim world. The concerns of generals, marshals, and mid-level officers had filtered through to the minister of defense, while party members and intellectuals had made their concerns clear to the leadership, sometimes appealing directly to the Politburo. The country's leaders were also becoming aware that since the secret operation had grown into a war, they would have ever-greater difficulty keeping it secret from the public. All of these factors encouraged Soviet leaders to go beyond the initial propaganda efforts of 1980 and look for other avenues to resolve the conflict—namely, the United Nations.

Toward a Diplomatic Solution

As the situation in Afghanistan became more challenging and the military lost confidence in its ability to stabilize the Karmal regime, the Soviet leadership began to consider a possible multilateral solution. Gradually the policy of supporting only direct DRA-Pakistan talks gave way to advocating a UN-mediated four-party discussion which would provide a legal framework for the Soviet withdrawal that began in 1988. Although there were still some key outstanding issues when Gorbachev came to power in March 1985, the accords had largely been prepared before the death of his predecessor, Konstantin Chernenko.

In the first half-year after the invasion, the Soviet leadership avoided any diplomacy that could, in the short term, limit its activities in Afghanistan. The Afghan Commission had come to believe that before the USSR could pull back its troops, much work would have to be done to build up the Karmal government. Karmal himself felt insecure and did not want to hear that Soviet troops might depart. In the words of Vasilii

Safronchuk, an adviser at the USSR's Kabul embassy in 1980–1982, Karmal stalled when it came to starting negotiations, "and Moscow helped him in this, so as to win some time to strengthen the new regime in Kabul and to increase the fighting ability of the army."[44] The only diplomatic initiatives that could be considered were ones that might enhance the legitimacy of the Karmal government—such as an offer by Cuban leader Fidel Castro to mediate in direct talks between the Pakistani government and the DRA.

Soviet leaders were willing to consider very limited diplomatic initiatives to help resolve the Afghan dilemma, realizing that the problem was not just the opposition within Afghanistan but also its support network, which included Pakistan, China, Saudi Arabia, and the United States. Although they rejected the possibility of involving the United States in these talks, they did accept Castro's offer to act as an intermediary in organizing talks between the DRA government and Pakistan. They also agreed that the United States and the Soviet Union could participate in the discussions, but at a latter stage. The priority was finding an agreement between Pakistan and Afghanistan.[45] Although this effort did not bring any immediate results, Pakistani president Zia ul-Haq did not reject negotiations outright, saying that while he could not recognize the DRA government, he welcomed Castro's mediation efforts.[46] The Afghanistan Commission confirmed this approach a month later, adding that similar efforts by other nonaligned countries would also be welcome.[47] Nevertheless, Soviet leaders rejected the UN's efforts at mediation, as well as various resolutions by the Islamic Conference and the European Community. At the same time, initiatives similar to Castro's made by other communist leaders, such as Romania's president Nicolae Ceauşescu, were unacceptable to Pakistan.[48]

The first indications that Moscow was becoming more interested in a diplomatic initiative came in the winter of 1980–1981. President Zia of Pakistan and the Soviet ambassador to Islamabad, Vitalii Smirnoff, held several discussions regarding the format of possible talks under the auspices of a UN representative. Moscow responded positively to the idea, but the initiative broke down because of a misunderstanding: the Soviets thought that Zia wanted a UN representative present, while in fact the Pakistani president had wanted a "special representative" who would organize the talks. The Soviet ambassador told UN Secretary General Kurt

Waldheim not to take this failed initiative as a sign that the USSR and Afghanistan were ready to accept a "trilateral" meeting.[49]

Although this initiative had gone nowhere, it opened the door for further UN efforts. When Waldheim traveled to Moscow in May 1981, he found Brezhnev and Gromyko more open to diplomacy and even to a more prominent role for the United Nations.[50] Gromyko said that the Secretary General's efforts should continue "at a cautious pace," but added that Moscow supported his efforts and was prepared to accept the participation of a special representative in the negotiating process. Gromyko added that Moscow "would cooperate with those efforts by advising the Afghan government to act likewise in this direction."[51] Javier Pérez de Cuéllar, acting as the personal representative of the UN Secretary General, found that both Kabul and Islamabad were also showing more interest in the possibility of a negotiated solution when he traveled there in August 1981. Though on his previous visits he had found the two sides unwilling to negotiate, he was now been able to secure agreement on an agenda for negotiations: withdrawal of foreign troops, noninterference by outside powers, guarantees of the agreed-upon provisions, and the rights of Afghan refugees.[52] However, when Pérez de Cuéllar mentioned this to a *New York Times* reporter, there was an angry reaction from Pakistan, which was still wary of admitting publicly that it was interested in such negotiations.[53] There would still be a considerable amount of such back-and-forth between the UN representative and Pakistani, Afghan, and Soviet officials before actual negotiations could get started.

In the meantime, Moscow showed increasing willingness to accept UN participation. Although earlier in the year Ustinov's letter questioning the wisdom of continued Soviet occupation had not even been considered by the Politburo, other members of the Afghanistan Commission were now becoming amenable to finding a diplomatic solution through negotiations. Kornienko recalls that in the autumn of 1981 the Ministry of Foreign Affairs prepared a memorandum—with support by Andropov and Ustinov—that proposed the acceptance of proximity talks between Afghanistan and Pakistan. The hope was that the resulting agreement would lead Pakistan to withdraw its support of the opposition in Afghanistan. The proposal was approved by the Politburo.[54]

After Pérez de Cuéllar was elected Secretary General, he appointed Diego Cordovez, an Ecuadorian lawyer and international official with twenty years of experience, as his personal representative on Afghanistan. Cordovez had already been involved in the preliminary efforts to start talks under Kurt Waldheim, but with the new appointment he would become the main UN official dealing with Afghanistan. After another trip to the area in April 1982, Cordovez was able to announce that talks would begin in Geneva on June 15, 1982.[55]

The press was not optimistic about Cordovez's mission, and his seemingly bottomless reserve of optimism and patience would earn him some ridicule over the years.[56] Indeed, for the time being, the new Soviet attitude toward negotiations was more a tactic than a profound change in strategy. Moscow saw the main purpose of the accords as a way for the Kabul government to gain legitimacy and strengthen its ability to fight the opposition. Nevertheless, the Soviet interest in negotiating was genuine. During the first round of negotiations, Moscow sent Stanislav Gavrilov, a senior official from the Soviet Ministry of Foreign Affairs (MID) who was also a specialist on the region, to act as a liaison with Cordovez. According to Vasilii Safronchuk, Gavrilov's superior at the MID, Gavrilov was known to have a low opinion of the Karmal regime and considered the invasion a tragic mistake.[57] That Moscow had sent him as a "minder" of the DRA representative confirms that the Soviets' attitude toward the Kabul government was changing.

The first round of the talks was held on June 16, 1982, in the Salon Français of the Palais des Nations in Geneva. With Cordovez acting as a go-between (the two sides never actually met in the same room), the Afghan and Pakistani foreign ministers made the first tentative moves toward an understanding on key issues: the withdrawal of Soviet troops and the cessation of all "interference" in Afghanistan. The results were minimal. As Cordovez put it, "The main significance of the talks was that they were held at all."[58]

There were technical as well as historical issues to be overcome. For example, Pakistan refused to admit that it was responsible for any interference. Shah Mohammed Dost—the Afghan foreign minister, who represented his country at the talks—then presented maps, provided by the Soviets, that showed the locations of *mujahadeen* camps on Pakistani territory. Yaqub Khan, the Pakistani foreign minister, told Cordovez that although such camps indeed existed, Pakistan could never admit this publicly. Cordovez eventually came up with a formula that bound both sides to stop interference, thus getting around an awkward problem.[59]

More serious was the issue of the Durand Line, the border between Pakistan and Afghanistan. The line had been established in 1893 by Sir Mortimer Durand, the foreign secretary of the Indian government, but it cut through what were traditionally Pashtun lands. Tribes had continued to move across the border as if it didn't exist. Disputes over the Durand Line had caused friction between Pakistan and Afghanistan, even leading to a major diplomatic crisis in 1953. Afghanistan refused to recognize the line as its proper border. Yaqub Khan argued that if interference were to cease, borders first had to be defined. The Afghan side disagreed, arguing that this issue should not be part of the discussions but should be resolved later on a bilateral basis.[60]

After the first round of talks (and several more discussions during the following UN session) Cordovez was able to produce a preliminary draft agreement. Although many of the details were left blank, the framework for a future accord had been established. Its four sections covered the withdrawal of troops, provisions on nonintervention and noninterference, a declaration of guarantees that such foreign noninvolvement would be maintained (with no mention of who would supply them), and a provision for the return of refugees.[61]

Kabul's participation in the talks and Moscow's support of the UN effort were not simply propaganda tools. Soviet leaders sincerely hoped that Cordovez's effort would help them find a way out. After Safronchuk briefed Gromyko on the talks at the end of June, the latter instructed him to find a solution to the Durand Line issue.[62] Similarly, when Cordovez and Pérez de Cuéllar came to Moscow in September 1982, they received some encouraging words from Brezhnev. Reading from a prepared statement and pausing for breath, the ailing leader told them that "as far as Afghanistan was concerned, the negotiations between Afghanistan and Pakistan had made a good beginning."[63]

In fact, the Soviet leadership was actively considering ways to withdraw from Afghanistan within a short period of time. Several weeks before Brezhnev's death, several senior officers serving in Afghanistan were summoned to Moscow to report to the Politburo. After several delays, the session finally took place on November 27, with Gromyko presiding. At the end of the meeting, Gromyko asked all the parties involved to draw up a plan for the withdrawal of Soviet troops from Afghanistan.[64]

For Moscow, the goal of any settlement was the preservation of the Karmal regime; its enthusiasm for the talks would be tepid so long as the regime itself remained weak. Brezhnev told Cordovez that the key issue remained the interference in Afghanistan by outside powers.[65] In some ways, the rhetoric had not changed. Pérez de Cuéllar writes in his memoirs that when he suggested that a regime change might be necessary in Afghanistan, the suggestion did not even get a response.[66] At the same time, Moscow was more than willing to press its clients on issues that it deemed of lesser importance, such as the Durand Line.

We cannot be sure if all of the senior leaders who sat on the Afghanistan Commission were equally enthusiastic about the UN effort, but there is evidence that each one individually was aware that a diplomatic track was necessary. Ponomarev, according to his subordinates, had been the most skeptical of the invasion from the beginning. Ustinov had already written, in 1981, that there was no military solution. The instructions that Gromyko gave to Safronchuk seem to suggest that he, too, saw the talks as important, notwithstanding his reserved attitude at the meeting with Pérez de Cuéllar and Cordovez. Kornienko, Gromyko's deputy, also confirms that in 1981 Gromyko had given his "blessing" to efforts for finding a diplomatic solution.[67]

Yurii Andropov's enthusiasm for the talks became clear once he became the General Secretary following Brezhnev's death in November 1982. Pérez de Cuéllar writes that Andropov was already hinting at his interest in a political settlement when the two spoke at Brezhnev's funeral.[68] Even more dramatic was Cordovez and Pérez de Cuéllar's meeting with Andropov in March 1983. While emphasizing that noninterference was still the key issue, Andropov complimented Cordovez on his efforts and told him that once there was an agreement on noninterference, all the other issues, including the withdrawal of troops, could be settled.[69] Setting aside his notes (something that Brezhnev did very rarely, particularly in his later years), Andropov said that, interference aside, the Soviet Union had no intention of keeping its troops in Afghanistan. Then, counting off on his fingers, he listed the difficulties the presence of Soviet troops had created: problems in relations with the United States, the Third World, and the Islamic world; negative effects on conditions within the USSR; and a consequent drain on the USSR's economy and society.[70]

Not surprisingly, some of the problems encountered at the first round of Geneva talks were settled by the end of the next round in June 1982. This included the issue of the Durand Line, where Cordovez was able to secure a text mutually acceptable to both the DRA and Pakistan.[71] The key issues that remained were noninterference, the rights of refugees, and the time frame for withdrawal of troops. Refugees were the only remaining major problem that did not require the direct involvement of the Soviet Union or the United States.[72] For the talks to be successful, however, it was now necessary to involve the two great powers directly, since they would have to act as guarantors.[73]

The Afghan situation had always been connected to the US-Soviet relationship, and this link again became apparent as issues that could be settled directly between Pakistan and Afghanistan were resolved. US-Soviet relations had been in near-crisis mode since the collapse of détente. The Carter Doctrine was enthusiastically adopted and even strengthened by the Reagan administration, which took office in January 1981. The United States showed little interest in the Secretary General's early efforts to negotiate a solution to the Afghan conflict, calling on the Soviet Union to withdraw troops as a precondition for an improvement in relations. US officials did not believe that Moscow's statements agreeing to a withdrawal were credible, and as a result they rebuffed most Soviet overtures. The crux of the problem was that Moscow did not want to commit to a time frame until Pakistan made a "formal commitment" to end interference. Similarly, Pakistan refused to commit to noninterference until Moscow agreed to set a date for the start of the withdrawal and accept a suitable time frame for its completion. Inevitably these issues would have to involve parallel negotiations with the United States, which was by far the biggest supplier of arms to the *mujahadeen* and in many ways led the multinational effort to support the resistance. In Andropov's words, the problem was not just "Pakistan's position. It is American imperialism that is giving us a fight. . . . We cannot retreat."[74] There could be no agreement with Pakistan until there was an accommodation with the United States. But the Americans did not believe that Moscow was really looking for a deal. As Charles Cogan, chief of the Near East and South Asia Division of the CIA's Directorate of Operations, explained: "We never considered that the Soviets would back out of Afghanistan and negotiate their way out. It didn't seem a credible thing for them to do, because we didn't think they were at all predisposed to do that. So naturally when we talked with the Pakistanis we pressed them always to continue the pressure."[75]

Moreover, for Reagan's administration even more than for Carter's, the

Afghan war was an opportunity to teach the Soviet Union a lesson about foreign intervention—a field where US Cold War hawks felt that their country had been repeatedly bested in the late 1970s. Aid to the *mujahadeen* was a linchpin of the "Reagan Doctrine," the effort to roll back Soviet influence in the Third World through support to anticommunist groups.[76] In addition, Reagan and many members of Congress saw support for the war in almost messianic terms, a sacred duty to help God-fearing freedom fighters to defeat arrogant, atheist oppressors.[77] "In Afghanistan," Reagan wrote in his memoirs, "the Soviets were attempting to subdue winds of freedom with a ruthlessness bordering on barbarity."[78] In the long term, this attitude made it especially difficult to reconcile Soviet and US positions on Afghanistan: even if both sides could agree on withdrawal and most technical issues, the idea of leaving behind a communist government was anathema to the US administration and many of its supporters.

Combined with skepticism about Soviet intentions, these attitudes encouraged administration officials to deflect Soviet overtures on Afghanistan. On a number of occasions, Dobrynin or Gromyko insisted that the Soviet Union was serious about withdrawal and wanted only an autonomous Afghanistan and the cessation of interference from Pakistan. The US secretary of state, Alexander Haig, would reply that the Kabul regime would not last without the support of Soviet troops. Typical was Haig's comment during a meeting with Gromyko in September 1981 at the US mission to the United Nations in New York: "[Haig] said that in all fairness he did not believe that under internationally acceptable circumstances with true self-determination this regime could survive five minutes. If the Soviet side accepted self-determination, then there was no issue. But if the present regime were to be retained as a façade, then the man [Karmal] would have his throat cut."[79] Although the official US position was still opposition to the Soviet invasion rather than to the Kabul regime, the administration was clearly looking not simply for a return to the status quo of December 1979, but for a full reversal. In this, they were in agreement with their Pakistani ally: Zia ul-Haq insisted that a political settlement within Afghanistan must accompany any broader international agreement.[80] And Haig refused to view the Afghan issue in isolation from other areas of confrontation, particularly in the Third World. He made it clear that he saw it as the most egregious example of Soviet interference in "areas of interest to the US."[81] A return to the status quo of 1970s détente was impossible, Haig said at another meeting with Gromyko in late September 1981. "Events such as Afghanistan were what SALT was impaled on. One could not deny the inter-relationship; one had to deal with it so as to improve East-West relations, because we had no desire for the current kind of situation."[82]

Haig did not completely discount the possibility of negotiating with the Soviets for a way out. Meeting with Gromyko again five days later, he said that he had reflected on their earlier discussion and that he saw some "convergences" and areas the two sides could work on in the future, including a phased withdrawal of Soviet troops, the broadening of the Kabul government's base, and steps to limit cross-border activities. Gromyko replied that he and Haig "might well have the opportunity to turn the key that opened the door."[83] They had reached an important mutual understanding: they both wanted a neutral, nonaligned, and sovereign Afghanistan. Gromyko likened this to a network of "US freeway interchanges."

Although Haig and Gromyko's discussions were followed up in Moscow by the US ambassador, Arthur Hartman, they nevertheless led nowhere. The two sides proved unable to establish the mutual trust they needed to negotiate on such a complicated and sensitive issue. In the months that followed, US officials saw increased Soviet activity in Afghanistan, while the Soviets saw increasing aid to the *mujahadeen*. From the US perspective, the imposition of martial law in Poland in December 1981 further poisoned the atmosphere. At their meetings in January and early July 1982, Haig and Gromyko were back to trading recriminations, although the latter did signal a willingness to pursue talks through the United Nations.[84]

Only with Haig's resignation in July 1982 did hope revive that US-Soviet talks could help untie the Afghan knot. But the appointment of a new secretary of state, more pragmatic than his predecessor and in principle more inclined to negotiate with the Soviet Union, was also accompanied by a growing willingness in Congress to increase aid to the *mujahadeen*. The Reagan administration, during its first two years, had supplied aid to the rebels at the level set under Carter: $30–40 million per year. From late 1982 onward, both the range of weapons and, ultimately, the sums allocated for aid began to increase.[85] In addition, the United States had launched a propaganda campaign alleging that Soviet forces were using biological weapons against the rebels, and helped put together a documentary claiming, among other things, that the Soviet military was using bombs disguised as toys to deliberately maim Afghan children.[86]

US-Soviet relations soon hit another rough spot with the downing of Korean Airlines Flight 007 in September 1983. The strong condemnation from the United States and the inept response by Soviet leaders meant that US-Soviet relations would remain at a high level of tension throughout the year. The US invasion of Grenada in October 1983 did not help, nor did the deployment of Pershing II missiles in Germany in November. One Politburo member described the international situation as "white hot, thoroughly white hot."[87] No wonder that when NATO's Able Archer military exercises simulating the outbreak of nuclear war with the USSR got underway in November, Soviet analysts thought that this meant a real attack was not far off.[88] Soviet intelligence, meanwhile, had apparently received word that Pakistani intransigence in negotiations was influenced by the United States.[89] Even under these severe conditions, however, Moscow continued to support negotiations, actively pressuring Afghan diplomats to cooperate with Cordovez's efforts, over Karmal's objections.[90]

By the time of Andropov's death in 1984, a draft of the agreement was nearly ready. Despite the problems in US-Soviet relations, Andropov had continued to maintain a strong interest in pursuing the Geneva talks and finding a diplomatic solution. Kornienko goes so far as to say that, had Andropov not fallen ill, the question of a timetable for withdrawal would have been solved by the end of 1983.[91] This is probably too optimistic, since the issue was closely linked to the question of guarantees. Its resolution depended on an improvement in US-Soviet relations. Nevertheless, Kornienko's comment confirms that in 1983 the Soviet leadership was looking for a way out through diplomacy. Questions regarding the timetable and noninterference would continue to be the main obstacles under Gorbachev, and would not be resolved until March 1988.

By the end of 1981, Soviet leaders had realized that their counterinsurgency strategy was not doing enough to prop up the Kabul regime. On the military front, they faced difficulties familiar to other regular armies fighting guerillas. They were often able to drive opposition fighters from a village or stronghold—but as soon as they pulled back or were ordered to another location, those fighters regrouped. They also found it impossible to close off the borders completely, meaning that supplies to the *mujahadeen* continued to flow from Pakistan.[92] The political side of the counterinsurgency strategy also failed. The Amin period had deepened the divide between the Khalq and Parcham factions of the PDPA. Karmal seemed to approach the question of unity only half-heartedly, preferring to purge Khalqi politicians and officers. As late as 1984, there was still very little PDPA presence outside the urban centers. At the same time, PDPA officials at all levels took very little initiative, preferring to let their Soviet advisers do the work.

Moscow came to support UN diplomatic efforts because it had begun to lose faith that its goals in Afghanistan could be achieved militarily; at the same time, Soviet leaders hoped that negotiations would lead to a cessation of arms supplies from Pakistan (crippling the Afghan insurgency) and to much broader recognition for the DRA government. After Moscow recognized the need to involve the United Nations, the diplomatic track achieved some success. By the time of Chernenko's death, the main bilateral issues between Pakistan and Afghanistan had been resolved. The remaining issues—a timetable for the withdrawal and a guarantee of noninterference—could not be resolved without some significant improvement in US-Soviet relations. The KAL 007 incident and the death of Andropov ended any chance of those relations improving in the short term.

By the end of 1983, the UN-sponsored negotiations were once again stalemated. Although several key issues had been resolved, there was no agreement on guarantees or on a timetable for the withdrawal. Nor would there be any movement on these issues under Konstantin Chernenko, the ailing leader who succeeded Andropov in February 1984 and stayed in power (if only nominally) until March 1985. Although negotiations continued through Cordovez, there was no progress on the key issues. A new Geneva round took place in August but was inconclusive.[93] With no movement in US-Soviet relations, it would have been very difficult to get past these two key issues.

3

GORBACHEV CONFRONTS AFGHANISTAN

By the time Mikhail Gorbachev came to power in March 1985, he had been a Politburo member since 1979 and had already earned a reputation with some Western observers as a reform-minded politician. He absorbed the ideas of the New Political Thinking and sided with those who supported Soviet initiatives to ease the Cold War confrontation. In formulating policy on Afghanistan, however, he had to balance his genuine desire to end the conflict with the immense legacy of support for the Third World that he and his team had inherited.

By October 1985, Gorbachev had decided to seek a withdrawal from Afghanistan and had won the support of the Soviet leadership to do so. Gorbachev was troubled not so much by the military and economic costs of the war or by the domestic political effects, although these were undoubtedly important, as he was by its potential to interfere with his broader reform projects, particularly his plan to seek a lessening of international tension. Like his predecessors, he was fully aware that the war had left the USSR isolated internationally, was difficult to explain to the population, dragged on indefinitely, and was disliked by many in the military and foreign-policy apparatus. His colleagues supported him because, even when they did not share his broader reform goals, they agreed that the continued presence of Soviet troops was unlikely to bring victory in Afghanistan.

But also like his predecessors, Gorbachev feared the consequences of a withdrawal, and believed that the USSR had to secure some sort of stability in Afghanistan before calling the troops home. The ideals of the New Political Thinking inevitably clashed with the reality of the Soviet Union's many commitments around the world and with its legitimacy in the eyes of its allies and even its own citizens. Gorbachev's failure to end the Soviet involvement before 1989 was a result of caution rooted in a belief that a perceived defeat in Afghanistan would entail a major loss of face, one that would not be well accepted by the Soviet Union's Third World allies. Gorbachev could bide his time because, like his predecessors, he was dealing with a limited war.

Old Commitments and New Political Thinking

Gorbachev's decision making on Afghanistan took place against the backdrop of a battle between the New Thinking in foreign affairs and the need to maintain Soviet power and prestige in the Third World. It was a battle taking place not only among Gorbachev's advisers and colleagues, but also within his own mind. A desire to end the intervention and seek a broad rapprochement with the United States and Europe had to be reconciled with the realities of ruling a great power and being the center of a chain of commitments, promises, and alliances that went to the core of the Soviet regime's ideology.

One obstacle to withdrawing Soviet troops from Afghanistan was the general climate of the Cold War in the early 1980s. Soon after Gorbachev came to power, he began looking for ways to reach out to the West, restart stalled arms control negotiations, and create a new basis for relations. This shift in foreign policy was conducted under the slogan of "New Political Thinking." The concept was less formal policy and more an emerging philosophy that, over time, came to characterize the conduct of foreign affairs under Gorbachev. It emphasized that confrontation was not inevitable and, crucially, that the Soviet Union could take major initiatives toward a lessening of tensions, thus differentiating its goals from the now-discredited calls for "peaceful coexistence" of the Brezhnev era.[1] Taking strong, often unexpected initiatives in negotiations with the United States on issues such as nuclear testing, force reduction, and regional problems became a key aspect of Gorbachev's style in foreign policy.

The roots of the New Thinking go back to the Khrushchev era and the political and intellectual thaw that began with the "secret speech" in 1956, delivered at the 20th Party Congress. Khrushchev's denunciation of Stalinism ushered in an era of relative intellectual freedom and debate that had not existed in the Soviet Union since the 1920s. New directions were taken in the fields of history and the social sciences, where many of the assumptions of the Stalinist period were challenged or pushed aside.[2] The period was also marked by the exposure of young social scientists and party members to Western ideas, scholarship, and ways of life. This took place through the creation of Soviet research institutes such as the Institute of World Economy and International Relations (IMEMO), the Institute of the Economy of the World Socialist System (IEMSS), and the Institute of the USA and Canada (ISKAN). Scholars at these institutes were given unprecedented access to Western scholarship, as well as an opportunity to study economic-policy experiments in Eastern-bloc countries such as Hungary and Czechoslovakia. Many of Gorbachev's advisers had also worked on the Prague-based journal *Problemy Mira i Sotsializma* (Problems of Peace and Socialism), where they were also exposed to "European" ideas while working alongside French and Italian communists.[3] Not surprisingly, representatives of this group, including Georgii Arbatov and Oleg Bogomolov, were the most vocal critics of the intervention and the biggest supporters of a speedy Soviet withdrawal.[4]

Although Gorbachev did not belong to this group of young intellectuals, he was influenced by the ideas they shared and developed. Gorbachev had attended Moscow State University in 1950–1955, at a time when Stalin was still revered for leading his country to victory in the Great Patriotic War. Still, Gorbachev made some important friendships with future reformers, including a leader of the "Prague Spring" movement, Zdenek Mlynar.[5] After graduation, he rose through the ranks of the Komsomol (the communist party's youth wing) and eventually of the party organization itself. Energetic, hardworking, and highly personable, he became, in 1970, first party secretary of Stavropol Krai, a largely agricultural area in the southwest of the USSR. Three years later he was made a member of the Central Committee of the CPSU. His steady rise, and his move to Moscow in 1978 as party secretary for agriculture, were facilitated by none other than Yurii Andropov, who was trying to assemble allies for his own struggle for power and his plans for reform. After Gorbachev moved to Moscow in 1978, he became acquainted with some of the reform-minded intellectuals working in the Central Committee and the institutes, including Georgii Arbatov and Anatolii Cherniaev. With Andropov's blessing, Gorbachev's interests quickly evolved beyond the agricultural sphere, his official domain, to larger questions of domestic and foreign policy.[6] Reform-minded thinkers like Georgii Shakhnazarov helped to develop the idea of a foreign policy guided by universal values and interests (rather than by the idea of class conflict), which

Gorbachev absorbed and eventually made a part of his own approach to foreign relations.[7]

Gorbachev's contacts with the New Thinkers in the years before he came to power not only had Andropov's blessing; they were in part ordered and supervised by the former KGB chairman. Aside from the informal discussions Gorbachev had with individuals like Arbatov and Aleksandr Iakovlev (one of the more radical reformers in his circle who had spent years in "diplomatic exile" as ambassador to Canada during the Brezhnev years), he also commissioned some 110 papers from them on various domestic and foreign policy issues, including Afghanistan, on Andropov's instructions. These papers and discussions provided the intellectual core of Gorbachev's reforms plans, and, as he put it in 1989, "formed the basis of the decisions of the April [1985] plenum and the first steps thereafter."[8]

Once Gorbachev was in power, he moved quickly to bring these advisers into the fold and raise the profile of their institutes. Many of them had been among the war's strongest opponents from the beginning, but during the later Brezhnev era they had lost much of their already limited ability to influence the decision-making process.[9] With Gorbachev in power, these New Thinkers were eager to reiterate their views on the war and make their recommendations. Soon after Gorbachev's election, Arbatov submitted a lengthy memorandum entitled "Toward a Revised Approach to Foreign Policy," which argued for an immediate withdrawal from Afghanistan. Other New Thinkers agreed. Anatolii Cherniaev, soon to become Gorbachev's foreign-policy aide but for the moment still working at the International Department, noted in his diary that if Gorbachev were to move quickly on Afghanistan, this would give him a major political boost: "Such an action would provide him with a moral and political platform from which he could later move mountains. It would be equivalent to Khrushchev's anti-Stalinist report at the 20th Congress. Not to mention the benefits the withdrawal would give us in foreign policy."[10] A good example of the New Thinking on Afghanistan comes from a memorandum submitted by a Gorbachev aide sometime in 1987:

> Our military presence in Afghanistan places an enormous financial burden on the USSR, and can lead to serious ideological consequences (the families of the dead); it damages our re-

> lations with the Muslim world, and gives the Americans an ideal opportunity to exhaust us by forcing us to wage an endless war. Of course, the withdrawal of troops and an agreement for some form of political settlement do not guarantee the survival of a socialist regime in that country. But however significant the survival of a socialist-oriented regime in that country is, in the end we will win. And the faster we leave that mousetrap, the better.[11]

To the New Thinkers, neither the survival of the PDPA regime nor the loss of prestige outweighed the costs of the war for the USSR. Moreover, they expected the withdrawal to pay dividends in the form of a better global image for their country and an improvement in their relations with the West.

As a philosophy, New Thinking was poised for a resurgence with Gorbachev's election. Not only was Gorbachev himself partial to the views of the New Thinkers, but many of their hard-line opponents had either left the scene or been pushed aside. The conservative champion of the military-industrial complex, Dmitrii Ustinov, had passed away in 1984. Viktor Grishin and Grigorii Romanov, both staunch conservatives, were removed from the leadership and sent into retirement during the July 1985 plenum, and Andrei Gromyko was asked to give up his job at the Foreign Ministry.

The New Thinking, with its emphasis on cooperation with the West, helped to lay the intellectual groundwork for the détente of the late 1960s and 1970s. But support for the Third World, particularly when it was part of a competition with the United States and China, was a legacy Gorbachev and his allies in power could not easily shake off. While those who laid the intellectual groundwork for the New Thinking at some distance from government could argue for a more radical reassessment of Soviet foreign policy, Gorbachev and Shevardnadze had to contend with the expectations of a large part of the communist world. During their first years in power, they brought little change to Moscow's relationship with Third World states. Even as the USSR underwent fundamental transformations in its domestic political and economic system, and in its relationship with the United States, China, and other former adversaries, it did not break its links with former Third World allies. Aid to the Third World continued well into 1990–1991, when the Soviet economy was nearing total collapse, and actually increased in the period 1987–1989.[12] (It did not help that even after Gorbachev hinted at a more conservative Soviet role in the Third World, Reagan maintained his commitment to supporting anticommunist forces, and spoke openly about rolling back Soviet influence there and telling "freedom fighters" in Afghanistan, Cambodia, Nicaragua, and Angola that they would have US support.)[13] Although the Soviet relationship with its Third World clients began to change after 1986, with the Soviet Union even helping to negotiate a peace deal in Angola, on the whole this change of pace was glacial, especially compared with the rapid downward spiral of the domestic economy. It was only after 1990, when the Soviet Union entered a serious economic crisis, that the foreign-aid budget decreased significantly.[14] Commitment to the Third World played an important role in Gorbachev and the Politburo's thinking about Afghanistan.

Little in Gorbachev's career had prepared him to deal with the Third World before he came to Moscow. Even after he joined the Politburo, his main focus was on agriculture, though with Andropov's encouragement he began to familiarize himself with other aspects of policy. His early policy toward the Third World was largely a continuation of Andropov's: he recognized that Soviet relations with Third World countries had to be reformed—that Moscow could no longer support leaders unconditionally, so long as they professed loyalty to the Soviet model. This was not about retreat, but about reinvigoration: the Soviet Union would help resolve intractable conflicts in places like Angola, improve relations with long-time foes such as China, and make a further push into Latin America. It would help to counter the indebtedness that more and more Third World countries were falling into. The Soviet Union would win in the Third World by reviving the appeal of socialism.[15]

To do so, of course, it would have to resolve the Afghan issue. But it would have to do this well; it could not afford to have its authority there undermined. The way potential Third World allies would react was a real concern to Gorbachev, particularly in this period, and not just something he brought up in Politburo meetings to highlight his reliability to conservatives. Nothing illustrates this better than his relationship with the Indian premier, Rajiv Gandhi. India, a leading nonaligned nation, was one of the most important countries for Gorbachev's Third World strategy, and Gorbachev "found it easier to talk with Gandhi than with Reagan or some other leaders. Their rapport was total and their discussions were genuinely frank." After the failure of his talks with Reagan at Reykjavik, he joined Gandhi in issuing the Delhi Declaration to outlaw nuclear weapons. India (along with the USSR and Afghanistan itself) also had the most to fear from a botched withdrawal from Afghanistan: a triumphant, resurgent Pakistan, confident that it had won an ideological victory and helped to defeat the mighty Red Army, would surely start causing more trouble for India. Gandhi was clear about where he stood, in his 1985 talks with Gorbachev: "It would be worse for all developing countries of the region if imperialism succeeded in strangling the revolution in Afghanistan."[16] The New Political Thinkers played an important role in convincing Gorbachev to withdraw from Afghanistan and helped him to shape the arguments for withdrawal. His own caution and his fear of a Soviet "failure" in Afghanistan, however, meant that he would spend several years trying to prevent the withdrawal from becoming a defeat before he brought the troops home.

A Decision, but Not a Plan

Mikhail Gorbachev came to power with a desire to end the Soviet intervention, but without any well-defined ideas about how to handle the Afghan situation, and he gave no immediate signs that he would seek a quick withdrawal. Although he gradually became more involved in decision making on Afghanistan, Gorbachev largely let the war run its course during his first year at the helm. Even after becoming convinced that disengagement from Afghanistan would require more direct intervention on his part, he moved cautiously, preferring to try every option available before finally giving up on helping the Afghan regime win the war.

In the spring of 1985, domestic political forces were aligned in favor of withdrawal, at least in principle. On the one hand, some of the key pro-interventionists were now gone and reform-minded leaders and advisers were gaining influence. On the other hand, some of the more conservative figures in the Politburo recognized that the war had become a quagmire and agreed that there was no military solution to the Afghan problem. The Afghanistan Commission, which under Gorbachev's predecessors had guided Afghan policy, still included Andrei Gromyko, but it also included the new minister of defense, Marshal Sokolov, and the new head of the KGB, Viktor Chebrikov. These men were conservatives, not New Thinkers. Yet Sokolov, who had served in Afghanistan, had become disillusioned with the war before he took over the defense portfolio from Ustinov. Chebrikov's views are unclear, but it is worth noting that he had come up in Andropov's footsteps, and thus probably shared Andropov's belief that the Afghan war was a mistake.[17]

Gorbachev had been a candidate Politburo member since 1979 and a full member since 1980. Although he was not privy to the work of the Afghan commission, he was certainly familiar with its reports and discussions of the problem at Politburo meetings, which he often chaired when Chernenko was ill. In fact, Gorbachev's name appears on records of Politburo discussions of Afghanistan going back to January 1980. And Cherniaev's diary mentions at least one Politburo meeting chaired by Gorbachev where the main subject of discussion was Afghanistan. Ustinov and Chebrikov, back in Moscow after talks with Karmal, painted a devastating picture of affairs there. The Afghan officer corps was still torn by the Khalq/Parcham split, almost half of the border with Pakistan was a "hole," and 80 percent of the territory was controlled by "bandits." Yet neither Ustinov nor Chebrikov, nor any of the other members of the Politburo, suggested that a radical change of course was necessary. With the asthmatic Chernenko constantly in the hospital and the political situation uncertain, matters were allowed to drift.[18]

Similarly, during Gorbachev's first months in power, there was little movement on Afghanistan. Though Gorbachev was not yet the radical reformer he would become later in the decade, he acted more decisively on some foreign-policy issues. At Chernenko's funeral he had already begun dismantling the so-called Brezhnev Doctrine, telling Eastern-bloc leaders that they could not rely on Soviet troops to keep them in power. By contrast, Gorbachev's comments during his first meeting with Karmal reflected some skepticism regarding the Afghan revolution but did not represent a radical break with policy. With foreign minister Andrei Gromyko, the last surviving interventionist, at his side, Gorbachev spoke about the need to expand the PDPA's base of support. The party had to attract a wider sector of the public, allowing for a "stabilization of the situation, consolidation of the revolution's victories, and resolution of some of the most difficult problems" facing the country. While he mentioned that Soviet troops "would not be in Afghanistan forever," he avoided specifics on how and under what circumstances they would be brought home.[19]

Although Gromyko did not disagree with Gorbachev's assessment, his few interjections during that meeting are revealing. Gromyko stressed the need for Soviet troops, but said nothing about internal reforms. Karmal said that while he and his party were working hard (as Lenin had taught them!) to improve their ties with the masses, closing off the borders with Pakistan and Iran was even more important, since this would "deliver a strong blow to the plans of American imperialists, Chinese hegemonists, Pakistani reactionaries, and other hostile powers." Gromyko agreed. Closing the borders, he said, "remained one of the most important problems."[20] Even before a policy was formulated, representatives of the old guard and the new were emphasizing different approaches.

Gorbachev's own thinking on Afghanistan evolved week by week and month by month. In 1985, Gorbachev was already looking to change course but was not sure what shape it would take. Supposedly on his first day in office he had already made a note for himself that Soviet troops had to leave Afghanistan, although the withdrawal would have to be done in stages.[21] Sometime in March or early April, Gorbachev requested a policy review from the sitting Afghanistan Commission, now composed of Marshal Sergei Sokolov (elevated to minister of defense after Ustinov's death), Andrei Gromyko, and Viktor Chebrikov, head of the KGB. The commission was told to look into "the consequences, pluses, and minuses of a withdrawal."[22] To Arbatov's call for a withdrawal, Gorbachev apparently replied that he was "thinking it over."[23] Even as he solicited advice and tried to formulate a new approach, he defended Soviet policy. In May 1985, Gorbachev told Italian prime minister Benedetto Craxi: "There is a certain process underway in this country, the point of which is to get rid of centuries-old backwardness. It is difficult to say when this will be completed."[24] Gorbachev defended the Soviet intervention: "Someone decided to interfere in Afghanistan's internal affairs. Under these conditions the USSR . . . introduced a limited contingent of its troops."[25]

For the next few months, the new General Secretary continued to discuss the Afghan problem, soliciting proposals from the likes of Arbatov, as well as from the Foreign Ministry and the military. Crucially, New Thinkers like Arbatov and the hard-boiled military men agreed that the war was hopeless. A report from General Valentin Varennikov, the head of the Ministry of Defense Operating Group in Afghanistan,[26] noted that military successes had no long-term effect on the opposition, which continued to grow. The DRA government failed at implementing the key counterinsurgency strategy of establishing a presence in areas cleared of guerrillas. As a result, "the combat actions for stabilizing the situation in the country can have only a temporary character. With time, the insurgents in these districts are capable of reestablishing lost positions."[27]

Gorbachev spent the summer of 1985 pondering the problem and soliciting advice on the Afghan problem. His aides and advisers noticed his growing willingness to confront the issue directly. One morning in the third week of June, Gorbachev even summoned Arbatov for a one-hour conversation that focused primarily on Afghanistan. Whereas previously he had told Arbatov that he was "thinking" about the Afghan problem, now Gorbachev said he agreed that a quick withdrawal was necessary.[28] By the fall, he was ready to start acting on the recommendations of New Thinkers and others who were urging withdrawal.

In October, Babrak Karmal was secretly called to Moscow for talks. Gorbachev put the problem in stark terms: the Afghan revolution had little popular support and needed to try a new approach quickly. He recommended a return "to free capitalism, to Afghan and Islamic values, to sharing power with oppositional and even currently hostile forces."[29] Gorbachev's advice to Karmal was not a complete departure from what his predecessors had advocated. Soviet leaders had long urged Afghans to adopt a slower approach that emphasized the establishment of political power over revolutionary rhetoric and programs. Gorbachev was going further than his predecessors, however. His advice to Karmal may have been the first time a Soviet leader had urged a client to turn to capitalism and religion, and it foreshadowed his own increasingly radical views after 1988. The record (though sketchy) of Gorbachev's conversation with Karmal shows a leader who had spent some time studying the problem and trying to understand the practical details of the situation.[30]

Gorbachev also made it clear that Soviet troops were not going to stay in Afghanistan much longer. In fact, in this conversation, Gorbachev set the first of several deadlines for the withdrawal: by summer 1986, Soviet troops would be out and the Afghans would have to "defend the revolution" themselves.[31] Najibullah, who was at the meeting, later said that Karmal's face went white when he heard this. Taken aback, he exclaimed, "If you leave now, next time you will have to send a million soldiers!"[32] Gorbachev told his Politburo colleagues that Karmal "in no way expected such a turn, was sure that we need Afghanistan more than he does, and was clearly expecting that we would be there for a long time, if not forever."[33] Gorbachev was learning the hard way that the Afghan communists would try to sabotage any withdrawal plan he could devise.

The next day, Gorbachev addressed the Politburo. He had clearly prepared for a decisive meeting that would result in a mandate for a new approach to the Afghan problem. After briefing his colleagues on the conversation with Karmal, he began reading aloud some of the letters that had been coming in to the Central Committee. Gorbachev not only cited letters concerning crippled soldiers and maternal grief over lost sons; he also quoted from letters that blamed the Soviet leadership directly: "The Politburo made a mistake, and it should be rectified, the sooner the better, because every day is taking lives." He concluded with a phrase that conveyed his disappointment in the Afghan leader: "With or without Karmal, we will follow this line firmly, which must in a very brief time lead to our withdrawal from Afghanistan." There was no objection to what Gorbachev said, and Marshal Sokolov, the defense minister, supported moving toward a withdrawal.[34]

Gorbachev's approach with his Politburo colleagues reflected his desire to establish a consensus on the Afghan problem. No doubt he had concerns about the reaction of some of his more conservative colleagues. Andrei Gromyko was still a supporter of continued intervention, as was evident at the meeting with Karmal in March. The Politburo's reaction, as recorded by Anatolii Dobrynin, seemed to justify Gorbachev's cautious approach: "There was no objection and no strong endorsement, but rather reluctant silent agreement."[35] By reading aloud letters from the public, Gorbachev was raising the "emotional tension" (as Cherniaev put it) and at the same time showing his colleagues that the public's tolerance for the war was limited.[36] Regardless of individual members' private concerns about the war, Gorbachev had to take the lead and form a consensus. He would need this consensus if critics later raised concerns about his handling of the problem.

Making the War Public

It was Gorbachev's initiative to open up press coverage that really began to change the way the war was perceived nationwide. For years, the Soviet press had mostly reported that the 40th Army was helping to build the revolution; only occasionally was it even acknowledged that Soviet troops were fighting. A *Pravda* editorial from February 1985 was typical of earlier coverage. Focusing on the US-led effort to supply the *mujahadeen* with arms, the editorial explains the USSR's involvement as vital not only for the success of the Afghan revolution, but also for the USSR's national interest. American policy was "one element of imperialism's anti-Soviet strategy," and the Afghan opposition was fighting on the side of the United States:

> If it succeeded in strangling the Afghan revolution and replacing the people's government in Kabul, the American generals, with the aid of the ringleaders of the Afghan counterrevolution, would not fail to take root in Afghanistan and provide themselves with bases there, and they would reconstitute the electronic reconnaissance centers that Washington lost in Iran. After all, one should not forget that the Soviet-Afghan border is almost 2,400 kilometers long. . . . It is known that the CIA and the Pentagon have long attached great importance to espionage concerning these regions.[37]

When the press reported on the war directly, it was to highlight the heroic deeds of individual soldiers.[38]

Gorbachev understood that if he was to pursue a qualitatively different course in Afghanistan, the Soviet public would need a better understanding of what was going on in the country. Coverage of the war began to change as early as the summer of 1985. In June, General Varennikov drafted a new set of guidelines that significantly expanded the aspects of the fighting which could be addressed in print and other media. For the most part the memorandum, approved at the end of July, focused on widening the reporting of individual heroic acts, as well as of minor combat engagements.[39] That same month, even before the final approval of Varennikov's memorandum, Soviet television for the first time showed footage of the fighting that was taking place in Afghanistan. For two and a half minutes Soviet viewers were shown young conscripts and burning vehicles.[40] Although it was only after the withdrawal that the Soviet media published real investigative journalism, bringing to light the truth about the intervention and the kind of war that Soviet soldiers had been fighting for more than a decade in Afghanistan, after 1985 the war was no longer hidden.

During the fall and winter of that year, the public read by far the most frank writing on the war to date. In August, Aleksandr Prokhanov, a journalist and novelist, published a long piece in *Literaturnaia Gazeta* entitled "Notes from an Armored Personnel Carrier."[41] While Prokhanov went on at length about the connection between the soldiers fighting in Afghanistan and the internationalists who had fought in Spain and the patriots who defeated the Nazis, he did not include the standard line about the importance of the revolution to the Afghan people. In fact, Prokhanov wrote, after five years "all illusions have disappeared."[42] Coming from Prokhanov, who had previously written some of the most enthusiastic propaganda about the war, these subtle changes were significant. Other, more explicit articles followed. Published letters and reports highlighted the difficulties of young veterans in readjusting to life back in the Soviet Union, and also pointed to the futility of the Soviet mission. One letter from a communist party official described his meeting with a young, recently demobilized veteran who had been "tormented by the fact that he was powerless to alleviate his comrades' suffering" and now was enraged by corruption at home.[43]

By 1985, then, increasing social pressure forced Moscow to reconsider how the war was presented to the public. Gorbachev's decision to bring it out in the open was part of his early effort to democratize the Soviet media, but it also served to make the war a more evident part of public discourse. Some of the negative effects of the war could now be freely discussed. This may also have been part of Gorbachev's effort to change the discourse on the war in the leadership: by appealing to public discontent, he could emphasize the need to withdraw. Crucially, however, from 1985 to the start of the withdrawal in 1988, Gorbachev was not in a position where public pressures or social problems were so severe that he was reacting to them in forming his Afghan policy. Public discussion was still controlled, and Soviet decision makers were not in the habit of responding to subtle changes in public opinion. Instead, Gorbachev was able to use such pressure as did exist to give himself freedom of maneuver and to preempt criticism from potential conservative critics.

While Gorbachev learned, debated, and formulated policy, the war in Afghanistan continued. Only a few months after he took over as General Secretary, the *mujahadeen* commander Jalaluddin Haqqani attempted one of the more daring attacks of the entire war. His goal was the capture of the city of Khost in southeastern Afghanistan, in Paktia Province, not far from the Pakistani border. Nearby was the Zhawar complex—a series of interconnected caves in which fighters and supporters trained and lived, and which reportedly included a hotel, medical facilities, repair shops, arms depots, and a mosque. It was also a distribution center for weapons smuggled from Pakistan. Taking Khost would have been a major blow to the morale of the Soviets and the DRA, but it would have had practical significance as well, allowing much easier movement and infiltration of fighters.

The ensuing series of battles again demonstrated what was wrong with the Soviet war effort—as well as with the *mujahadeen*. Haqqani's fighters successfully overran a number of DRA and Soviet positions, but these were retaken by Soviet reinforcements. Mohammad Yousaf, the coordinator of the resistance effort for the ISI, tried to achieve some coordination between various commanders, but without success. Meanwhile, DRA and Soviet commanders decided to push the fighters out of the Zhawar complex once and for all. At the end of the summer, a DRA-led force moved against the *mujahadeen*; after it took several towns, the offensive stalled: they had seriously underestimated the defensive capability of the resistance, which included heavy artillery and tanks brought over and manned by DRA deserters. An attempt the following spring to destroy the cave complex almost ended in complete disaster: troops were accidentally dropped inside the Pakistani border; others were killed by *mujahadeen* ambushes on landing positions. Finally, after fifty-seven days of fighting, the Soviet-DRA force was able to take the complex. Sappers were sent in to mine it, but they had to work quickly: fear of a counterattack meant that they were out within five hours. The time was insufficient. Although they destroyed some entrances, the *mujahadeen* were able to defuse or safely detonate a number of the mines, and many of the weapons stored there survived unharmed.[44]

Whether the fiascos at Zhawar had any direct impact on Gorbachev's thinking is unclear, but it seems that he formed his impressions primarily on general trends and developments, on the assessments of his advisers, and on his interactions with Afghan leaders. Although there has been some speculation that he gave the military a two-year window to win the war, this seems unlikely. None of the Soviet sources support this theory, and Liakhovskii, for one, denies that any such directive was given. Nor did Gorbachev authorize a "surge," as has been suggested. But during his first year in power there was indeed an increase in the number of troops in Afghanistan, primarily special forces (SpetsNaz), possibly as a result of a decision made before March 1985 and reflecting Soviet adjustments to the demands of the war. But during this period, there was also an "Afghanization" of the war. The DRA forces were now to play the leading role in operations, even if in practice Soviet forces still had to come to the rescue on most occasions, as in the Zhawar battles.[45]

In 1980, the most senior members of the Politburo had sought a Central Committee plenum to ratify their decision to send troops into Afghanistan and keep them there to fight on the side of the government. The plenum had ratified the decision unanimously, as expected, giving the Politburo a "mandate" to continue its Afghan policy. Now that a new direction was being set for Afghan policy, it would have to be ratified by the party as well. This "ratification" took place at the 27th Party Congress in February–March 1986. Apparently, serious attention was given to placing the need to withdraw troops from Afghanistan on the political agenda. Eduard Shevardnadze, who barely mentioned the war in his first memoir, wrote that the topic was in the early drafts of the Congress's Political Report, but had been removed, presumably at the insistence of more hard-line advisers or Politburo members.[46] In any case, Gorbachev, in his keynote speech, still called the Afghanistan war a "bleeding wound," thus telling the assembled delegates, the nation, and the world that the Soviet leadership saw the war as a drain on the country.[47] We may never know who pushed for the withdrawal item to be removed from the Political Report; but in light of what *was* said, this is not of great importance. It may well have been a tactical decision: by making the point in his speech but not putting it on the agenda officially, Gorbachev indicated that ending the war was now a priority but also left room for the USSR to do this on its own terms.

The winter of 1985–1986 was a critical point in Gorbachev's evolution as a leader and in his conceptualization of reform in both the domestic and international spheres. Publicly, Gorbachev often still used the language of the Brezhnev period—for example, referring to denunciations of Stalinism as "foreign propaganda."[48] Following the disappointing meeting with Reagan at Geneva, Gorbachev and his advisers sought new approaches. Ultimately, they rejected the "two camps" formula, which highlighted the confrontational nature of relations between capitalist-

and socialist-inspired states. Instead, they articulated a new approach with a focus on integrity and interdependence.

Despite skepticism from some of Gorbachev's Politburo colleagues, these new foreign-policy ideas became crucial components of his report to the 27th Party Congress in February 1986.[49] Still, there was no firm decision on how Afghan policy should be conducted in order to make a Soviet withdrawal possible. A withdrawal decision *in principle*, which the Politburo had approved in October 1985, was neither a strategy nor a plan in practice. In fact, while Gorbachev clearly wanted to move Soviet policy toward a withdrawal, he did not yet have any particular scheme in mind. More generally, Gorbachev operated on the assumption that the Soviet Union needed to withdraw from Afghanistan without "losing face."[50] In April 1986, two months after labeling the war a "bleeding wound," Gorbachev told a special Politburo meeting that a poorly conceived withdrawal from Afghanistan would do great harm to Soviet relations with its client states. In the presence of Varennikov and the ambassador to Kabul, Fikriat Tabeev, Gorbachev said he believed "we must under no circumstances just clear out from Afghanistan, or we will damage our relations with a large number of foreign friends."[51]

With the weight of the Soviet Union's commitments to the communist world on his shoulders, Gorbachev feared acting precipitously. Perhaps nothing highlights this better than the statement he made at a February 1987 Politburo meeting: "We could leave quickly . . . and blame everything on the previous leadership, which planned everything. But we can't do that. They're worried in India, they're worried in Africa. They think that this will be a blow to the authority of the Soviet Union in the national-liberation movement. Imperialism, they say, if it wins in Afghanistan, will go on the offensive."[52]

Despite a general consensus that the war had to end, the Soviet effort in Afghanistan was too closely associated with Soviet efforts elsewhere in the Third World and with its reputation as a guarantor of friendly regimes. Bringing home the troops was not in itself a problem; but if the withdrawal was followed by a collapse of the Kabul government or a *mujahadeen* victory, there could be manifold consequences for the prestige of the Soviet military, for the USSR's reputation as an economic benefactor, and for its effectiveness as a political role model. Though by early 1986 Gorbachev's foreign-policy views were already a significant departure from those of his predecessors, he was not about to unravel the entire fabric of Soviet foreign policy and its attendant myths. He sought a breakthrough with the United States, but specifically omitted from the agenda any discussion of "regional" issues and "solidarity" with "fighters for independence."[53] And Gorbachev knew that if the worst were to happen as a result of a withdrawal he had initiated, it would be great fodder for the conservatives, who were already growing suspicious of his turns in foreign policy. Meanwhile, the war remained within boundaries that the Soviet state could tolerate. The wound may have been bleeding, but the patient was not in mortal danger from massive blood loss.

The myriad difficulties Soviet leaders encountered as a result of their introduction of troops into Afghanistan had an effect that went beyond their decision making with regard to that unfortunate country. Not only had the intervention brought East-West confrontation to an uncomfortable level and complicated relations with Third World allies; it also threatened to embarrass the Soviet military and its ability to defend socialism abroad. When the Polish crisis erupted in 1980, Soviet leaders felt strongly that they could not afford to "lose" Poland, but they were far less confident about using Warsaw Pact forces to crush the "counterrevolution."[54] Soviet leaders began to see the costs of intervention as outweighing the benefits.[55] There was also a clear sense of hangover from the support of Marxist and quasi-Marxist regimes in the 1970s and a feeling of frustration regarding Soviet aid efforts, as evidenced by Andropov's statement in 1983: "It is one thing to proclaim socialism as one's goal, and it is quite another to build it." The fate of progressive states, he went on, depended on "work by their own people and a correct policy on the part of their leadership."[56]

Gorbachev's policies in 1985–1987 continued to be shaped largely by the initiatives undertaken in 1980–1985, although they were modified and pursued with renewed intensity. This is not surprising, since there were few alternatives to these policies which did not involve abandoning the PDPA regime and accepting a government dominated by the opposition. Such a scenario would have been too big of a blow to Soviet prestige and to the interests of conservative leaders such as Brezhnev, Andropov, Gromyko, and Ustinov. Paradoxically, the realization that the PDPA continued to be weak seemed to draw Soviet leaders deeper into the quagmire, as they assumed the Afghan government's functions on the military and political levels.

Gorbachev's desire to bring Soviet troops home stemmed not just from considerations about the war's military, economic, and social costs, but also from the influence of New Thinking party members—intellectuals who were opposed to the war from the beginning. Even before he became General Secretary, Gorbachev was the Politburo's point man for interacting with these intellectuals and soliciting reform ideas from them. They, in turn, shared their belief that the war was a mistake—and did so formally, through policy papers, and informally, when they were invited to brief Gorbachev or offer advice. These New Thinkers had a view of international relations quite different from the confrontational one that had dominated since the fall of détente at the end of 1979, and they saw the Afghanistan war as the worst of the Brezhnev era's foreign-policy mistakes. Gorbachev seems to have agreed. Furthermore, by the time Gorbachev came to power, the war had been dragging on for five years without any significant result.

Gorbachev and the team he assembled around him inherited not only a system but an entire legacy of foreign-policy making, with its attendant history, myths, and commitments. Whatever may be said about Gorbachev with hindsight, he was intent neither on ending the Soviet empire nor on dismantling the Soviet state. A reformer rather than a revolutionary by nature, he was greatly worried about the consequences of his actions. Thus, even while calling the war a "bleeding wound," Gorbachev worried about the effects on relations with client states if the USSR was seen as suffering defeat in Afghanistan. Crucially, while Gorbachev viewed the war as a tragic mistake, this was not how he viewed the broader Soviet policy of "solidarity" with Third World states and national-liberation movements. He believed in Moscow's obligations and the importance of maintaining its role as guarantor. Though his attitude would start to change in mid-1987, in 1985 and 1986 he saw himself as carrying the mantle of Soviet leadership in the Third World as this relationship had evolved since the 1950s.

Such ideas, which some consider sentimental, might have mattered little if the military, economic, and social costs of the war had been so staggering that Gorbachev and his colleagues would have recognized the need to cut their losses and bring the troops home. Yet the Soviet intervention in Afghanistan never put that sort of pressure on the Soviet state. The fact that the war dragged on for years after Gorbachev and the Politburo agreed that it was time to leave is explained by the legacies of old commitments, as well as by the nature of the war. A "bleeding wound" it may have been, but the flow came from a small vein of a large animal.

Leonid Brezhnev embracing Nur Muhammad Taraki, December 1978. Brezhnev would later complain that Taraki's execution in September 1979 was a personal insult, since the whole world had seen them embrace only a few weeks earlier. His aides and colleagues took this as a cue that the General Secretary was coming around to the idea of intervention. TASS photo.

Babrak Karmal, January 1980, looking less than confident at his first televised press conference after he took power with Soviet help. A photo of Taraki is in the top-left corner. TASS photo.

Karmal's visit to Moscow, October 1980. From left to right: Nikolai Tikhonov (chairman of the USSR Council of Ministers), Sultan Ali Keshtmand, Leonid Brezhnev, Babrak Karmal, Anahita Ratzebad, and Andrei Gromyko. Ratzebad, who had studied in Chicago, was a doctor by training, and was the only female member of the Politburo. She and Karmal were reportedly lovers; she too was pushed out of the leadership by Najibullah in 1986. Photo courtesy of Vladimir Snegirev.

Ahmad Shah Massoud, "Lion of the Panjsher Valley," at a graduation ceremony for volunteer fighters, 1986. Next to him is Ahmet Muslim Hayat. Massoud built not only an effective fighting force, but a remarkable network of hospitals, schools, and services. Soviet officers and diplomats marveled that what they had been trying to help the PDPA government achieve, Massoud seemed to be doing on his own, and better. Photo courtesy of Colonel Ahmad Muslem Hayat.

Somber faces: The first meeting of Babrak Karmal and Mikhail Gorbachev. To Gorbachev's right is foreign minister Andrei Gromyko and foreign-policy adviser Andrei Aleksandrov-Agentov. GFA photo.

All smiles: Mohammed Najibullah with Mikhail and Raisa Gorbachev. Najibullah, the Soviets felt (or hoped), understood "the need for perestroika, for moving toward the New Thinking, for making and fulfilling decisions directed at settling the Afghan problem through political means." GFA photo.

Tashkent, April 1988. Putting a brave face on the upcoming withdrawal. Left to right: Rafik Nishanov (Uzbek first secretary), Mohammed Najibullah, foreign-policy adviser Anatolii Cherniaev, foreign minister Eduard Shevardnadze, Mikhail Gorbachev, and an interpreter. "Even in the harshest, most difficult circumstances, even under conditions of strict control—in any situation, we will provide you with arms," Gorbachev promised. GFA photo.

Eduard Shevardnadze signing the Geneva Accords. They were highly unfavorable to the Kabul regime, and he knew it. Shevardnadze later wrote, "I knew that we would not lessen our political efforts for a peaceful settlement in Afghanistan, but still I could not rid myself of a sense of personal guilt toward my friends." GFA photo.

Yulii Vorontsov meeting with Pakistani prime minister Benazir Bhutto, 1989. Vorontsov worked tirelessly both before and after the Geneva Accords to bring peace to Afghanistan. He was "one of the real heroes of this story," according to his colleague Nikolai Kozyrev. TASS photo.

Soviet tanks returning home over the Friendship Bridge, which linked Soviet Uzbekistan with Afghanistan. TASS photo.

Boris Gromov, last commander of the 40th Army, talking to reporters on the Soviet side of the Friendship Bridge on February 15, 1989. His son is in the background. No one from the Soviet Politburo came to greet the returning soldiers, a fact they did not forget. TASS photo.

4

THE NATIONAL RECONCILIATION CAMPAIGN

By the spring of 1986, there was a consensus in the Soviet leadership that the military involvement in Afghanistan had to end. In October 1985 the Politburo had agreed with Gorbachev that it was time to pull out, and at the party congress in February 1986 the Soviet leader had publicly called the war a "bleeding wound." Yet the withdrawal started only in May 1988. During the period in between, the USSR made its last major effort to shore up the Kabul regime under its tutelage and with massive Soviet aid. The war was not going so badly that an imminent withdrawal was necessary, particularly when such a withdrawal would be balanced against the potential political costs. But somehow the war had to be brought to a close without the potential negative consequences. As one of the General Secretary's aides put it, "Gorbachev hoped that he could right the mistakes of his predecessors while paying no political price."[1]

Such a goal meant, first of all, that the Soviet involvement could not end in *porazhenie*. This Russian word literally means "defeat" but suggests something even more shattering than a military reversal—it connotes humiliation. In fact, the Soviet position at this time differed little from Yurii Andropov's stance three years earlier. First, outside interference had to stop, a condition that required an agreement with the United States and Pakistan. Second, there had to be some international recognition of the DRA regime, even if the character of that regime would be allowed to change, at least within certain bounds. Third, the DRA regime had to outlast the Soviet troop presence. At no point could the USSR be perceived as bowing to international pressure.

Not only were Moscow's demands and preconditions for withdrawal in 1986 similar to those issued in 1983; its approach to the Afghan problem was similar as well. Beginning in 1985, there was renewed emphasis on the four-party Geneva talks, which had previously stalled. From 1987,

Moscow also undertook a renewed effort to stabilize the Kabul government, and to make it more self-reliant and acceptable to the Afghans. The new initiative, baptized "National Reconciliation," stressed reaching out to the clergy and to peasants, winning over elements of the opposition, and using Afghan traditions to secure legitimacy for the government. It was a case, as Yulii Vorontsov later recalled, of "doing everything possible to withdraw in good order."[2]

This is not to say that the 1985–1988 period was just a repeat of 1982–1983. Indeed, while the initiatives, plans, and hopes were similar in the broad outline, there were significant differences. National Reconciliation attempted to go further and deeper, in terms of transforming the Kabul government, than anything undertaken in the years 1980–1985, and Moscow's efforts in Geneva showed more flexibility than similar efforts under Andropov. There was an effort, too, to change the *manner* of Soviet involvement in Afghanistan—an effort aimed at improving the DRA government's self-reliance. And of course Moscow initiated a key personnel change with a good deal of support within Afghanistan, replacing Babrak Karmal with Mohammed Najibullah.

Ultimately, these efforts proved insufficient, and in 1988 Moscow accepted a withdrawal on terms much less favorable than those it had initially sought. One cannot understand either the progress of the war in the years 1985–1988, or the progress of events after the withdrawal, without looking at Moscow's hopes and efforts with regard to what it could accomplish in Afghanistan.

Exit Karmal, Enter Najib

Moscow's goal in Afghanistan had always been to create a government stable enough to function without Soviet troops. Under Gorbachev, this effort was renewed with greater force. When it became clear that few changes could be expected while Babrak Karmal remained in power, the Politburo sought to replace him. The replacement of Karmal became the first of several steps Moscow took in 1985–1987 with the aim of reforming the DRA government. It gave Soviet leaders hope that a friendly regime could still be preserved in Afghanistan, and that their influence could still to engineer an internal reconciliation.

Soviet leaders had decided to install Karmal in 1979 because they saw him as a moderate and conciliatory figure, someone who could help to heal the divisions caused by Taraki and Amin. They were to be disappointed. For all his oratorical skills, Karmal proved indecisive, unable to push either his party or his country in any particular direction. Moscow should have known better: a profile compiled in December 1979 by Soviet military intelligence, the GRU, noted that Karmal was "a skillful orator" but "emotional, with a tendency toward generalization rather than concrete analysis. He has a poor command of economic problems, and is interested only in their general outline."[3] It was also rumored that Karmal drank rather heavily, despite warnings from his Soviet advisers. In an interview with the weekly magazine *Ogonek* in 1989, Varennikov described him as "a demagogue of the highest order" who "did not deserve the trust either of his own colleagues, or of his people, or of our advisers."[4] It was inevitable that after 1985, as Moscow looked ever more urgently for a way out of the Afghan quagmire, Karmal would be pushed aside.

It is not clear at what point Moscow first began seriously to consider replacing Karmal. One interesting clue comes from the archives of the UN Secretary General: documents reveal that in 1982 the possibility of removing Karmal was discussed by Pérez de Cuéllar and foreign minister Andrei Gromyko, although it is not clear how supportive the latter was of the idea.[5] In any case, Soviet leaders had almost certainly considered getting rid of Karmal before 1985. Indeed, it is possible that by 1983 or thereabouts, there was general agreement that Karmal would need to be replaced at some point.[6]

Ultimately, however, it was Gorbachev who pushed for Karmal's dismissal. Gorbachev's disapproval of Karmal was clear in October 1985, when the two had what seems to have been their first substantive discussion.[7] Although in the fall of 1985 Karmal did take some steps to implement Gorbachev's recommendations, preparations were soon underway to replace him with Mohammed Najibullah. Najib, as he preferred to be known, was the son of a wealthy civil servant; he was a graduate of the prestigious Habibia College and of the medical faculty of Kabul University.[8] Like Karmal, he had joined the party at its creation, and soon took charge of an underground university organization. After practicing medicine for several years, he turned to party work full time, joining the Central Committee in 1977, at the party's "reunification." During the period of Khalqi domination, he stayed in Yugoslavia, returning to Kabul in December 1979 to lead the State Information Service, or KhAD, the secret police. In 1981 he entered the Politburo of the PDPA, where he took charge of a commission on tribal relations, and also became part of the Defense Council.[9] J. N. Dixit, the Indian ambassador, first met Najibullah in 1982 and found a man "in his late thirties" with a "heavy mane of hair, [and a] black-brown moustache." Dixit noted that Najib "exuded confidence, and facts and figures were at his fingertips. [Najib] conveyed the impression of being efficient, competent, assertive, and alert."[10] Najib had earned the nickname "Ox" for both his physical strength and his forceful personality. In 1986 the burden and the honor of trying to run the party and the government of Afghanistan would pass to him; it was the "Ox," Moscow hoped, who would help build the exit road for Soviet troops.

Najib had caught the eye of Soviet agents in Kabul as well as of leaders in Moscow even before the intervention, and they liked what they saw when he took over KhAD. They were impressed, in part, by his ability to establish links with Pashtun tribal leaders. He was well known to Ustinov, Andropov, and Ponomarev, all of whom thought highly of him for similar reasons. Apparently, at some point prior to 1983 they had already consulted former Khalqis, and had come to the conclusion that the only viable replacement for Karmal, when the time came, would be Najib.[11]

Najib was neither the sole candidate to replace Karmal nor obviously the best choice. A GRU report from April 1986 pointed out that many in the PDPA leadership preferred Assadullah Sarwari, a Khalqi, who was head of KhAD's predecessor agency under Taraki. Sarwari, the report suggested, was a better candidate because he could unite the party as well as balance the interests of the Pashtuns with those of the Tajiks, the Uzbeks, and others. Najib, by contrast, was a Pashtun nationalist, unlikely to reach out to non-Pashtuns.[12] Some consideration was also given to General Abdul Kadir, the military leader who had sided with the PDPA in April 1978 and made the coup possible.[13]

It is not clear how much debate took place regarding Karmal's possible replacement, but in all likelihood it was minimal. The KGB had been grooming Najib for the leadership for several years; he was both a Pashtun and a Parchamist, and was believed to have excellent organizational skills. His promotion in November 1985 to secretary of the PDPA Central Committee, where his portfolio included managing relations with Pashtun tribes, is further testament to the Soviets' determination to remove Karmal and to the fact that Najib was chosen as a replacement quickly and without much debate.[14]

In March 1986, when Babrak Karmal was invited to Moscow for discussions and for medical treatment, Soviet leaders tried to convince him that he had to step down—that his health was poor and he should make room for someone younger. Apparently, there was some awkwardness later, when Soviet doctors treating Karmal told him he was in fine health.[15] Nevertheless, Karmal saw that he had little room to maneuver in Moscow and so did his best to be allowed to return to Kabul. He claimed to understand the situation and promised to act differently and pay greater heed to Soviet recommendations.[16]

Karmal was allowed to return to Kabul on the condition that he step down as head of the party, remaining only as chairman of the Revolutionary Committee. Soviet leaders suspected that he would try to cling on to both positions, and so Vladimir Kriuchkov, then head of intelligence at the KGB, was sent after him. According to the available evidence, Karmal indeed proved obstinate.[17] In a lengthy and emotional monologue, he professed undying loyalty to Soviet leaders. A true Muslim, he explained, honored God, his prophet, and the four righteous caliphs. He proclaimed that his feelings toward the Soviet Union and its leaders were close to this veneration; it was a principal foundation of his life. Kriuchkov persisted, stressing that Karmal's own colleagues wanted him out of the way. Finally Kriuchkov left, asking permission to return the next day.[18] A few hours later, the Afghan ministers of defense and of the security services came to see Karmal and insisted he had to relinquish one of his posts. Finally realizing he had no cards left to play, Karmal gave in.[19]

Following the 18th Plenum of the PDPA, held in May 1986, Najib became the chairman of the PDPA Politburo as well as of the Defense Council. The plenary session "granted Comrade Babrak Karmal's request that he be relieved of his duties as General Secretary of the PDPA Central Committee for health reasons."[20] But Karmal still had influence and a good deal of support, and he used these to undermine Najib's position. Prior to the second party conference, in 1986, Karmal's supporters spread rumors that Najibullah would be removed and Karmal reinstated as General Secretary. The source of these rumors was the Ministry of State Security, or WAD, which contained, thanks to an earlier Soviet initiative, a core of people devoted to Karmal. WAD disliked Najibullah because he supposedly aimed to "clean up" the agency, whereas under Karmal its agents had a free hand.[21]

Karmal's maneuvering did not meet with much sympathy in Moscow. At first, Soviet leaders preferred to proceed cautiously in replacing him. They must have realized that Karmal had significant support from sections of the party. Furthermore, they had been his steadfast supporters for six years; if they abandoned him too quickly now, their patronage of other Afghan leaders would count for less. In September 1986 Gorbachev directed Yulii Vorontsov, the new Soviet ambassador, to ask Najib not to rush ahead with the firing of Karmal.[22] All the while, Najib was earning the respect and even loyalty of his Soviet interlocutors. A note submitted to the Politburo in November 1986 by Dobrynin, Sokolov, Shevardnadze, and Chebrikov said: "It is clear that he is disposed toward finding real approaches to the problem [of National Reconciliation]. He needs our support in this, especially since the PDPA is far from unanimous in accepting the idea of reconciliation." For his part, Najibullah urged Moscow to support him in ousting Karmal completely, claiming that Karmal had "abandoned party and government work" and occupied himself with "fault finding" and "speaking out against National Reconciliation."[23]

As far as Soviet support was concerned, Najib did not need to worry. The Soviets had decided that he was their man, and no longer felt there was any reason to keep Karmal around. At a meeting on November 13, 1986, the Politburo decided that Najibullah had to be given more leeway to act independently and that Karmal had to be removed completely.[24] Several Politburo members spoke out in favor of Najib, including Gromyko, KGB chief Chebrikov, Shevardnadze, and Vorontsov. While Gromyko spoke in favor of letting Karmal serve as a figurehead, others, including Dobrynin, said that Karmal had to go.[25] The November PDPA plenum relieved Karmal of his last remaining post. He soon left for Moscow, where he was given a state-owned apartment and a dacha. Although he returned to Afghanistan in 1989, he never regained influence, and died in Moscow in 1996.

Although there may have been other candidates, Najib was ultimately acceptable to everyone in Moscow as well as to Soviet officers and advisers in Afghanistan. His original patrons may have been Soviet KGB agents, but Najib quickly won over the rest of the Soviet leadership and the senior officials working on Afghanistan, who thought he had the willpower and the capacity to carry out National Reconciliation. Even the military tended to see him as a highly capable organizer with whom they could work.[26] The Politburo protocol assessing his December 1986 visit to

Moscow noted that Najib could be expected to begin a major restructuring of his party and government: "The ideas expressed oriented the Afghan government toward *perestroika* in the shortest possible term in all spheres of party, government, military, political, and economic activity—toward a decisive turn in practical policy and in the direction of achieving national reconciliation in Afghanistan." Najib was the person to carry this out, since he had shown his understanding of "the need for *perestroika,* for moving toward New Thinking, and for making and fulfilling decisions directed at settling the Afghan problem through political means."[27] Politburo members felt that in Najib they had found the right man for the job. "He creates a very good impression," Shevardnadze said after meeting him in January 1987. "He is taking the initiative in his own hands."[28]

Najib impressed Soviet leaders as a serious, pragmatic politician who understood the Soviet desire and intention to disengage from Afghanistan. With Najib at the helm, Soviet leaders wanted to give their efforts some more time to bear fruit. Over time, this faith in Najib came to have a complicated and even dangerous effect on Soviet perceptions. Vadim Kirpichenko, deputy chief of the KGB First Directorate, later wrote that Najibullah's success in establishing more control within Kabul and some sectors of the government led them to believe that they had found a solution that could be replicated everywhere in Afghanistan: "Faith in Najibullah and in the dependability of his security organs created illusions on the part of the KGB leadership. . . . These dangerous illusions, the unwillingness to look truth in the face, delayed the withdrawal of Soviet troops by several years."[29] For Shevardnadze and Vladimir Kriuchkov, director of intelligence and later KGB chief, faith in and commitment to Najibullah came to define Moscow's relationship in Afghanistan.

Gorbachev supported changing the Afghan leadership because he believed that this would improve the situation in the country and allow him to bring the troops home. Trying to draw lessons from Afghan history, Moscow backed an ethnic Pashtun who, they hoped, would emerge as a friendly yet independent strong man.[30] Najib was a communist, but undogmatic in his political views and fiercely proud of his Pashtun identity. His succession reflected the crucial role the KGB continued to play in the Soviets' Afghan policy, one that would become even more obvious after the withdrawal (as we will see in Chapter 5). Later events would show that whatever his qualities as a leader (and these were considerable), he was a far from ideal candidate to lead Afghanistan in National Reconciliation. His desire to hold on to power and his distrust of non-Pashtun politicians led him to reject alliances and truces favored by his Soviet advisers. With the support of the KGB and key figures in Moscow, however, Najib learned he could usually get his way.

National Reconciliation

Gorbachev and many of his colleagues still believed that they could create a successful government in Afghanistan as long as the regime gave up any effort to transform Afghanistan along Marxist lines and focused instead on gaining legitimacy through traditional Afghan institutions. In 1987, Moscow began changing its approach to counterinsurgency in Afghanistan. Previously, the emphasis had been on winning over the population through economic incentives and establishing party and government influence in the cities and countryside. The new initiative continued that policy but placed a much greater emphasis on pacification through winning over rebel commanders.

The Policy of National Reconciliation (PNR) was planned and written by Soviet advisers, with representatives of the military, the foreign ministry, and the KGB all taking part.[31] National Reconciliation was largely what Moscow had been preaching, and what the PDPA had theoretically been doing, since 1980. The principles of what the Soviets urged Karmal to do and what they urged Najib to do were quite similar. Gorbachev's injunctions to Karmal in October 1985, cited in the previous chapter, were part of a continuing leitmotif: "Widen your social base. Learn, at last, to pursue a dialogue with the tribes, to use the particularities [of the situation]. Try to get the support of the clergy. Give up the leftist bent in economics. Learn to organize the support of the private sector."[32] Soviet agencies had a role to play as well. The KGB, the Ministry of the Interior, and the Ministry of Defense, for example, were tasked with engaging with frontier tribes to help close the border with Pakistan. Both the KGB and the Ministry of Foreign Affairs were to take part in encouraging opposition groups to come over to the government side. Overall, a dozen Soviet ministries, party committees, and government offices were drafted to take part in National Reconciliation.[33]

Certain elements of Moscow's policy were clearly a case of "old wine in new bottles." Among these, one can count a massive aid package intended to give Najib a boost as he embarked on reforms. Moscow coordinated aid with a number of East European countries, with the biggest aid package, of course, coming from the USSR itself.[34] In February 1987, the Politburo agreed to provide 950 million rubles' worth of gratis aid, more than the USSR had ever given to any one country.[35] This was as much a political as an economic move. Najib needed to show supporters and rivals in the DRA that the Soviet Union would support him.[36] In March, Gorbachev also promised Najib that after the withdrawal had taken place, absolutely all of the military infrastructure would be handed over to the DRA armed forces to help them protect the "independence and sovereignty" of Afghanistan. Some of the economic aid would even go to helping Najib develop the private sector, which was considered a necessary precondition for the success of National Reconciliation.[37]

Moscow also sought to make Afghan politicians more independent and to change the way Soviet advisers there operated. As discussed in Chapter 1, Moscow's policy had fallen into a trap. On the one hand, the presence of advisers seemed to discourage Afghan officials from taking any initiative, either in decision making or in policy execution. Readers will remember that an assessment of the PDPA from 1983 noted that this tendency reached the highest levels of the party.[38] On the other hand, the PDPA's seeming impotence only encouraged Moscow to send more advisers.

Its is difficult to determine the extent to which advisers really "ran" the country, but it is clear that their presence was felt everywhere and often gave Afghan politicians an excuse to shirk responsibility. Najibullah later described what he called a typical meeting of the Afghan council of ministers: "We sit down at the table. Each minister comes with his own [Soviet] adviser. The meeting begins, the discussion becomes heated, and gradually the advisers come closer and closer to the table. So accordingly our people move away, and eventually only the advisers are left at the table."[39]

After Najibullah replaced Karmal, the presence of Soviet advisers could still be felt everywhere within the Afghan government. Even as the USSR sought to shake off accusations that it had become yet another empire trying to dominate vulnerable Third World states, the omnipresent Soviet aid workers inadvertently gave the impression that a colonial occupation was under way. Soviet advisers, Vorontsov commented, were "everywhere, absolutely everywhere. It was the worst sort of colonial politics. Terrible."[40] Even senior Soviet figures, such as ambassador Fikriat Tabeev, were often guilty of imperiousness in their dealings with Afghans. A party man who had spent twenty years as the head of the Tatar Autonomous Soviet Republic, Tabeev had been appointed in 1979 in part because of his Muslim background.[41] Over the years, he had begun acting as a "governor general," and had apparently been telling the newly promoted Najib, "I made you a General Secretary."[42]

Such behavior was inconsistent with Moscow's emphasis on Afghan self-reliance, and Moscow sought a different figure to coordinate its new approach to the Afghan conflict. In late 1986, Tabeev was replaced by Yulii Vorontsov, a career diplomat with experience in South Asia.[43] Many advisers were withdrawn in 1986, and the Soviets tried to change the way relations with Afghans at every level were conducted. Experience showed that this would be far from easy, and the attitudes of Soviet advisers as well as the fractiousness among institutions continued to be a problem. In May 1986, the Politburo had discussed removing Tabeev so that relations with the Afghan leadership could be placed on a different footing; in June and July, the Politburo moved to recall many of the advisers and specialists.[44]

Vorontsov was more than just Moscow's ambassador to Afghanistan in this period—he was effectively the most senior diplomat coordinating the effort to find a diplomatic solution. He worked tirelessly, trying to come up with initiatives and concessions that would be acceptable to his bosses in Moscow, to his counterparts in Afghanistan, Pakistan, and the United States, and to the opposition leaders. He was, as his colleague Nikolai Kozyrev put it, "one of the true heroes of this story." Vorontsov's first task, though, was to coordinate the work of the various institutions involved in Afghanistan: the KGB, the military, the Foreign Ministry, and the political advisers. Gorbachev was aware that there was a difficult relationship among these institutions and that their recommendations often conflicted. Vorontsov was given a "mandate" to coordinate their work and to provide the Politburo with recommendations on which all parties could agree. It was tough work—the representatives of these various institutions could become quite forceful in their disagreements. But the "mandate from the General Secretary" helped.[45]

Even though National Reconciliation was supposed to attract opposition leaders to the regime, the character of Soviet and government propaganda changed little. Throughout the war, the Soviet army had an uneasy relationship with the population and the DRA. In August 1987, seven months after the start of PNR, Colonel Shershnev, who had argued for a change in the Soviet army's approach as early as 1984, called together the entire propaganda division of the main political directorate (GlavPU) of the Afghan army and suggested asking the "higher-ups" to stop calling the opposition "a band of killers," "mercenaries of imperialism," "skull-bashers," and so forth. Only in February 1988 did a response come from GlavPU, saying that the terms of reference would be changed. "From now on, the counterrevolution would be called 'the opposition' instead of 'armed band of hirelings.' . . . It took over a year after the proclamation of PNR for our military leadership to start calling the *mujahadeen* 'the opposition,' which is what it was."[46] For Soviet officers who needed to lead their soldiers into battle, this was a war to be fought and an enemy to be crushed, not a political project to unite former adversaries.

Soviet military officers sometimes seemed unwilling to alter their strategies radically to bring them in line with the principles of Moscow's new policy. Throughout the war, some officers had actively engaged with the population in the vicinity of their deployment, establishing ties with local leaders, providing medicines and other products to villagers, and mediating in disputes with soldiers.[47] Much depended on individual officers, however, and they did not always see themselves as being part of the broad political effort: the military's job was to fight, while the diplomats, advisers, and KGB officers could worry about negotiating with insurgents and winning over the population. In 1987, Vladimir Plastun, an orientalist working for the GRU, tried to convince Colonel General Vladimir Vostrov that military attacks on Kandahar Province were counterproductive. The general chided Plastun: "To hell with national reconciliation. Warriors receive medals on their chest and stars on their epaulettes and money not for reconciliation but for conducting combat operations. This is something that you, expert, did not understand!"[48] Although not all officers took such a hawkish approach, and at least some took the initiative in trying to make peace with the insurgents, the military often seemed reluctant to do its part in political work. In March 1988, its most senior officer (and the one who put in the greatest effort to arrange a reconciliation between the DRA government and its opponents), General Valentin Varennikov, complained, "Our army is not just a warrior with a sword. It is a political warrior. . . . Over the last year, meetings between Soviet and

Afghan soldiers have ceased, as have those of Soviet soldiers with the population."[49]

Yet if mid-ranking officers sometimes failed to grasp the political significance of their operations, senior commanders were very active in efforts to co-opt certain opposition commanders. Along with Soviet diplomats, they tried to help the National Reconciliation process along by opening their own talks with leaders of the opposition. One leader that Soviet military chiefs and some diplomats thought was particularly promising was Ahmad Shah Massoud, the Tajik "Lion of the Panjsher Valley."[50] On several occasions the military had been successful in concluding a cease-fire with him, which in turn had kept the northern area fairly quiet from 1981 to 1983. Following the announcement of National Reconciliation, Massoud sent a feeler to representatives of the DRA government in the Panjsher Valley, but the attempts at talks collapsed when Kabul insisted that he lay down his arms. Nevertheless, Massoud instructed his forces to maintain a virtual cease-fire and to undertake no offensive action.[51] In October 1987, General Varennikov succeeded in opening discussions with the Tajik commander, although these collapsed when news of the contact became public.[52]

Efforts to work with Massoud were not limited to the military. The Soviet ambassador to Kabul, Yulii Vorontsov, who had been instructed to develop such contacts as part of his contribution to National Reconciliation, studied Massoud's biography and speeches and concluded that his support was essential. Vorontsov was particularly impressed that Massoud seemed interested in, and capable of, organizing a development program—building schools, hospitals, and roads in his area. Vorontsov wrote to Moscow suggesting that the Soviet Union could offer to help Massoud financially in developing his region if he would ally with Kabul; Moscow sent a favorable response. Vorontsov was able to arrange a meeting with Massoud, but it was sabotaged at the last minute. Apparently, an Afghan Air Force jet had bombed the Panjsher Valley, causing Massoud to call off the meeting. The Soviet command had not been informed of the attack—once again, Kabul had sabotaged a Soviet effort to open a dialogue with Massoud.[53] Nevertheless, Soviet attempts to contact Massoud continued, and would become particularly important in 1988 when the 40th Army was withdrawing through Massoud's territory.

It was not only the advisers and the military who had trouble radically changing the course of Soviet-Afghan relations. At every level, there were vestiges of the "colonial" approach that had taken root since 1980. Although Tabeev had been removed, and there was much talk about making Najib "more independent," it was much harder to cleanse the relationship of all manifestations of imperialism. General Ziarmal, chief of the political directorate of the DRA army, complained in January 1988 that the practice of having Najibullah meet the Soviet minister of foreign affairs at Kabul airport on his visits there only underlined the colonial nature of their relationship. Perhaps this seemed like a minor point, Ziarmal said, but "in the eyes of international opinion it makes Afghanistan a satellite of the USSR." Ziarmal went on to complain that even the Soviet press did not take this point seriously: "Shevardnadze, for us, is a comrade; but Najib, for Shevardnadze, is 'Your Excellency the President of the Republic of Afghanistan.' The Soviet press needs to speak about him as a president, not the General Secretary of the PDPA."[54] Soviet failures in this regard only compounded the enormous difficulties in Najib's attempts to create legitimacy for his regime.

Within Afghanistan, National Reconciliation was launched with great fanfare but produced few immediate results. In January 1987, the government announced a cease-fire and invited its opponents to negotiate. A general amnesty was also announced later that month. Officially, at least, the government was rolling back from its revolutionary aims on a broad front. Restrictions on private commerce, which Soviet officials had urged the Afghans to loosen, were lifted. The word "Democratic" was removed from the country's name, which henceforth became the Republic of Afghanistan (RA). Officials spoke increasingly of the government's Islamic character. In mid-1987, a new law on political parties was passed, essentially granting legal status to all parties that did not take up arms against the state or against other parties. (Although parties did not have to share the PDPA's ideology, they did have to agree on the importance of "strengthening the historical friendship with the USSR.") In November 1987, a Loya Jirga (a grand council convened for major national events and decisions) approved the country's new constitution, and in April of the following year elections took place for the new parliament. The actual gains resulting from these measures are hard to gauge; most of the armed opposition, certainly, did not take these initiatives seriously, and it boycotted the April 1988 elections.[55]

Critics of National Reconciliation argued that it was just a propaganda campaign, and they were not wrong, at least as far as its early months are concerned. Najibullah would ultimately prove capable of reorienting the government quite dramatically, but for the moment he was hampered by divisions in his party and by cadres who were afraid they would quickly lose authority if they began to share power. The fractiousness of the PDPA by this point was not limited to the Khalq/Parcham split, but included intrafaction groupings that formed around the more senior members. Aside from the Khalq/Parcham divide, there were groups loyal to individual leaders: "Karmalists," "Nurovists," "Wakilists," and "Keshtmandists." There were also those loyal to the deposed Hafizullah Amin, although most of these were in prison until 1988. Many party leaders viewed National Reconciliation negatively because they believed their patrons' position would not be secure in a coalition government.[56] Indeed, the party, reconstituted as a single entity only through Soviet efforts, would probably have fallen apart without Soviet influence. As late as December 1987, the head of Afghan army propaganda said that a formal split would almost be preferable, with each wing choosing its own leader and Najibullah remaining as president.[57] Najib, meanwhile, had to face the residual support for Karmal in the party and the government. On the day of Najib's election, teachers and students had marched in support of Karmal in Kabul. Pro-Karmal sentiment was strong even in KhAD, despite the fact that it was Najib who headed the organization before taking the reins of the country. At the July 10, 1986, Central Committee meeting, Najib added forty-four loyal members to the committee, but it took over two years to purge the Politburo and the Central Committee fully of Karmal's allies.[58]

In 1986 and particularly in 1987, as Gorbachev's domestic reforms seemed to stall, a new leitmotif entered Politburo meetings. Gorbachev and the reformers were frustrated that even as new approaches were adopted at the top of the Soviet hierarchy, the wheels of change ground much more slowly closer to the bottom, at the level of lower party organs, ministries, and enterprises. There were similar difficulties in reforming the way Afghan policy was conducted. Lack of unity in the PDPA made it difficult to guide the party toward a new path. Soviet military and political advisers also seemed slow to adopt the new approach that was supposed to bolster Afghan independence. Concurrently, the Policy of National Reconciliation did not seem to have much support among lower-level PDPA cadres. In the view of Soviet officials, the Afghan communists avoided participating in PNR, and did not direct the resources provided by Moscow to the population. At the same time, afraid of being punished for their party activity, governors, mayors, and chairmen used whatever clan, family, or tribal ties they had to ensure their own safety.[59] Over a year after Najib had taken the top party post and Moscow had begun to recall its advisers, Shevardnadze was forced to admit: "In the work of our advisers [in Afghanistan], despite our instructions and our discussions at the Politburo, there has been no turning point."[60]

In backing Najib, the Soviet government hoped it had found a strong leader who would take charge of the party and the government and would not be seen by his own people as a puppet of Moscow. Yet Najib's draft speech to the 19th PDPA Plenum, written with much input from Soviet advisers, was full of references to Gorbachev's "advice, recommendation, and approval." As Cherniaev wrote to Gorbachev, this was contradictory to what the Soviet Union was trying to do in Afghanistan, since "one of the factors facilitating a decisive change in Afghanistan and the widening of the social base in Afghanistan is a demonstration of 'sovereignty' in the decisions taken by the new leadership and in the policies it conducts."[61]

The presence and protection of Soviet troops allowed the PDPA leaders to move slowly with regard to National Reconciliation. The pullback, as well as disagreements among Soviet officials regarding what National Reconciliation meant and how it should be implemented, deprived Moscow of leverage over its clients. Najibullah found ways to sabotage Soviet-led outreach when he felt it suited his interests. After the Soviets withdrew, the PDPA took much more courageous steps in terms of opening up the government and society, establishing links with tribal leaders, and shedding its communist image—all of which helped the RA government survive into 1992. By mid-1987, Soviet leaders realized that the Policy of National Reconciliation would not be able to guarantee the survival of a friendly regime in the near term, and that they would have to look to other solutions if they wanted the troops to come home.

In 1983, Andropov had spoken of scaling back Soviet efforts in the Third World, insisting that those countries had to depend not on Soviet aid and advice but on "work by their own people, and a correct policy on the part of their leadership."[62] In Afghanistan, this principle was finally becoming policy. Soviet advisers were being pulled back and the regime was encouraged to be more independent. Rather than relying on Soviet "state building," the stabilization effort would now rely on an Afghan leader who would, it was hoped, patch together a government with enough non-PDPA and opposition support to be legitimate.

A Return to Diplomacy

The diplomatic effort to end the Afghan war, an effort that had shown great promise in 1982–1983 but had largely stalled by the time of Andropov's death, was restarted several months after Gorbachev came to power. By the end of the summer, negotiators had made significant headway, but once again the lack of a Soviet-American dialogue limited further progress. Yet Soviet diplomacy was not focused merely on the UN-sponsored Geneva talks. In the context of National Reconciliation, the KGB, the military, and the Foreign Ministry all became proactive in making contacts with opposition leaders, with the ultimate goal of persuading them to join the government.

Gorbachev turned to the Geneva process because restarting the talks was the first logical step toward untying the Afghan knot. Indeed, the first Geneva round of the Gorbachev era was quite promising. Prior to the June 1985 talks, Soviet interlocutors managed to convince the Afghans to affirm that withdrawal, and a cutoff of aid would take place simultaneously. In May, Moscow had sent strong signals to Javier Pérez de Cuéllar that the Soviet Union was interested in a new round. The Soviet-Afghan side also made it clear that it was prepared to link withdrawal formally to the entire package, "which they had refused to admit for the last two years, i.e., since Mr. Andropov's exit from the political scene."[63] In Geneva, the negotiators were also able to produce an agreement on international guarantees.[64]

On the whole, however, diplomacy continued to be difficult. While Moscow seemed more interested in a dialogue, the United States was skeptical that there had been a real change in policy. The US undersecretary of state, Michael Armacost, told UN mediator Diego Cordovez that he had noticed "a lot of hints of a different Soviet style" but not a change in substance. Both Moscow and Kabul continued to insist that no real progress on the actual issue of withdrawal could take place until the Pakistani government was ready to sit down with the DRA.[65]

By August 1985, the texts of the first three instruments of the Geneva Accords—including ones on interference, the return of refugees, and international guarantees—had been completed. The main outstanding issues were the status of the Kabul government, which Pakistan did not want to legitimize, and the time frame for the withdrawal.[66] Yet the August round of the Geneva talks began with a standoff about the format. Shah Mohammed Dost, the Afghan foreign minister, insisted that if Pakistan continued to refuse direct talks, he was prepared to wait "two or three years" before continuing negotiations. Ultimately, the entire round was restricted to discussion of format issues and a review of the negotiations—a limitation that greatly frustrated the UN and Pakistani officials.[67] Throughout the fall, Cordovez and Pérez de Cuéllar tried to convince Kabul and Moscow that having a procedural impasse at that point would discredit the entire Geneva process. In New York, foreign minister Mohammed Dost continued to press for direct talks between Kabul and Pakistan, arguing that Islamabad's refusal showed "that Pakistan did not really want a settlement."[68]

US-Soviet discussions on Afghanistan moved no faster, despite the fact that both US president Ronald Reagan and Mikhail Gorbachev were interested in improving relations between their countries. Gorbachev did not yet trust the Americans sufficiently for him to engage with them directly and overcome the hurdles that had stalled the Geneva talks during Andropov's tenure. Equally important was that Moscow was just embarking on the process that would become National Reconciliation and an overhaul of its efforts within Afghanistan. As long as Soviet officials held out hope that their new approach would pay dividends, radical diplomatic moves were out of the question. At the same time, Soviet interest in the Geneva process was genuine, and throughout 1986 Moscow continued to look for ways to push it along. The February 1986 "bleeding wound" comment was only the first of a series of calculated comments and decisions made to signal Moscow's willingness to seek a diplomatic solution.

With the Geneva talks stalled, Moscow wanted to send another signal—one that showed the Soviet Union was serious about disengagement but that didn't weaken its bargaining position. At the end of June 1986, following another unsuccessful Geneva round, the Politburo considered the possibility of withdrawing some troops. So far, Gorbachev noted, the effort to find a political settlement had not been working; the United States seemed uninterested and was "picking on every little thing." Maybe the best course would be to withdraw 5,000–10,000 troops.[69] Two weeks later the Politburo approved a proposal to remove 8,000, to show "that the USSR is not going to stay in Afghanistan and did not want 'access to warm waters'"—a reference to Western critics who claimed that the intervention was the first step toward Russia's fulfilling its long-held ambition to obtain access to the Persian Gulf.[70] To underline the importance of this signal, Gorbachev announced the withdrawal during a major speech in Vladivostok in July.[71] In an address often cited by policy makers and historians as a turning point in Soviet relations with East Asia, Gorbachev highlighted the Soviet desire to get out of Afghanistan and the decision to withdraw six regiments "by agreement with the DRA": "In taking such a serious step—which we previously communicated to interested governments, including Pakistan—the Soviet Union aims to speed up the political settlement, to give it another push. This also comes [from our desire] that those who organize and carry out the military intervention against the DRA will understand and evaluate this step correctly. The response should be the end of such interference."[72]

By the middle of 1986, all the textual issues regarding the Geneva Accords had been resolved and discussions regarding a time frame for the withdrawal were becoming more concrete. Having supported a change of leadership and promoted National Reconciliation, however, Moscow began pursuing a separate diplomatic track that was at once a part of and separate from the Geneva talks being carried on by Cordovez. As Riaz Khan, the Pakistani negotiator, put it, "The Soviets linked withdrawal to progress in achieving political reconciliation inside Afghanistan, thus forcing the negotiating process into an entirely new arena."[73]

Soviet diplomacy was not limited to the Geneva process. Moscow placed increasing emphasis on opening a dialogue directly with Pakistan, a likely player in any effort at reconciliation. At the end of September 1986, the Politburo discussed conducting a "secret exchange of ideas" with Pakistan on the possibility of expanding the Kabul government by inviting émigrés to participate.[74] These conversations intensified toward the end of 1986. At the end of September, Shevardnadze and Pakistani foreign minister Yaqub Khan met in New York. That same month, the Pakistani Foreign Ministry took note of a statement made in an informal setting by Georgii Arbatov, to the effect that Najib would have to accept refugees and some *mujahadeen* in his government. The most productive discussions between Pakistani and Soviet officials began in December, when Yulii Vorontsov invited foreign secretary Abdul Sattar to Moscow.

Vorontsov explained the forthcoming National Reconciliation plan and told Sattar that the Soviet Union had firmly decided to withdraw, but also that a "cooling off" period was required to avoid bloodshed. During that period, various Afghan parties could observe a cease-fire and engage in discussions.[75]

Throughout 1987, Moscow tried to use its new ties with Pakistan to promote National Reconciliation, all the while insisting that the final success of the Geneva Accords was linked to progress on the "second track." At the end of January 1987, deputy foreign minister Anatolii Kovalev traveled to Islamabad to meet with senior officials and the president, Zia ul-Haq. The conversations that took place were unprecedented, and they also revealed the gap that remained between Soviet and Pakistani positions.[76] Kovalev said that "winds of change were sweeping across the Soviet Union," which made it all the more imperative to untie the Afghan knot. But he also continued to insist that National Reconciliation had to be linked to withdrawal, and that this had to take place under Najibullah. The Pakistani idea—a neutral interim government not headed by anyone associated with the present regime—was unacceptable to Moscow.[77]

The next month, Pakistani officials traveled to Moscow, meeting there with Shevardnadze and others. Prior to their departure, Pakistan had been able to obtain some negotiating positions from a reluctant alliance group. While the parties to the alliance made it clear that they were willing to pursue a nonaligned Islamic foreign policy and provide safe passage during the withdrawal, they also insisted on direct negotiations with the USSR. The Soviets, sensitive to Afghan opposition to such direct talks, which legitimized the opposition's claim to power, could not agree to this. Although Shevardnadze did not completely reject the Pakistani idea of bringing back the king to head a government, the meetings in Moscow did not move much beyond what had already become clear the previous month in Pakistan: the Soviet Union was not prepared to see a coalition government headed by someone from outside the PDPA. And although they were prepared to have the king return as a figurehead, Soviet officials could not fathom replacing Najibullah.[78]

Moscow's engagement with Pakistan was hampered by President Zia's commitment to the removal of the Kabul regime on the one hand and the disunity of the *mujahadeen* on the other. Zia had a personal, ideological commitment to the resistance and to Afghan policy in general. Earlier in the war, he had seen the *mujahadeen* resistance as way to block or discourage further Soviet encroachment. With a withdrawal imminent, he became seriously worried about the consequences of a power vacuum; hence his sudden and unexpected insistence, in the winter of 1988, on the formation of a government *prior to* the signing of the Geneva Accords.[79] Yet despite playing a crucial role in organizing international support for the resistance, the hosting of their groups on Pakistani soil, and crucial logistical support, even Zia could not dictate a settlement to the *mujahadeen*. Zia's proposal for a national-reconciliation formula that would divide power equally among the government, the *mujahadeen*, and monarchist groups greatly impressed Vorontsov and supposedly even his superiors in Moscow.[80] It failed to impress Yunus Khalis, the leader of the Alliance of Seven, who made it clear that the formula might be Zia's but that it was not the formula of the resistance. As for the leaders who succeeded Zia, they had virtually no clout with the resistance leaders; anyway, they were too busy trying to stay in power to establish a coherent line, let alone dictate to the ISI or the resistance groups.[81]

The Peshawar-based Alliance of Seven was no government-in-exile, but rather a motley group of charismatic warlords with varying motivations, personal and corporate ambitions, and networks of fighters. Early in the war, Zia had decided not to encourage a true government-in-exile, for fear that Soviet leaders might counter by setting up governments-in-exile to challenge Pakistani rule in restless areas like the North-West Frontier Province.[82] Zia's reluctance in this regard only encouraged the existing tendency for disunity and rivalry; cooperation would have perhaps been wiser. In any case, the group's disunity only contributed to the difficulty of working toward the formation of a government; even their Pakistani interlocutors found that they could get the resistance leaders to adopt a united position only after sustained pressure.[83] The fact that the ISI encouraged these leaders to think of improving their military capability while the Foreign Office tried to elicit negotiating positions from them could not have helped.[84] Even when the Alliance Interim Government was cobbled together prior to the signing of the Geneva Accords, it never functioned as a proper government-in-exile, and the individual leaders who were its members continued pursuing strategies primarily for their personal advantage.

Moscow's efforts to reopen diplomacy in the summer of 1985 produced a level of dialogue on Afghanistan not seen during the entire previous course of the war. The decision to engage with Pakistan was particularly important in this regard. Nevertheless, Moscow's commitment to leaving behind a PDPA-led, and increasingly Najib-led, government in Kabul meant that an enormous gap remained between the Soviet position and the US-Pakistani one. The reluctance of the United States to respond positively to Soviet "signals," such as the "bleeding wound" speech or the withdrawal of some troops in 1986, contributed to the stalemate—but this was not the only reason that a diplomatic solution was still beyond reach. Equally important was that Moscow was in the midst of reorienting its approach in Afghanistan, and still had hopes that with a new leader in Kabul a domestic reconciliation could still be achieved. These hopes were not realized.

Disappointment

By early 1987, Soviet leaders had started to realize that the situation in Afghanistan was worse than they had thought when they replaced Karmal. The economy was ruined, Najib was still isolated in the Kabul government, and a withdrawal was no closer than it had been in October 1985. The term discussed for withdrawal in January and February 1987—two years—was almost the same as the one Gorbachev had been calling for a year and a half *earlier*, after Karmal's visit. The year 1987 became a crucial turning point in the war, because month by month Moscow acknowledged the gravity of the Afghan problem—a severity from which, to some extent, it had been shielded before. Even more important, Soviet leaders were becoming aware that their 1986 plans for saving the DRA government were insufficient.

Shevardnadze and Dobrynin traveled to Kabul soon after Najib's visit to Moscow in December 1986. Upon their return, Shevardnadze gave the Politburo a devastating report on Soviet activity there:

> Of friendly feeling toward the Soviet people, which had existed in Afghanistan for decades, little remains. Many people have died, and not all of them were bandits. Not one problem has been solved in favor of the peasantry. In essence, we fought against the peasantry. The state apparatus is functioning poorly.

> Our advice and help is ineffective. . . . Everything that we have done and are doing is incompatible with the moral character of our country.[85]

What is striking about these exchanges is how different they are from the discussions that took place before 1985. Euphemisms and comments regarding "significant progress despite certain difficulties" were noticeably absent. Shevardnadze was not the only one to speak up in this way. Marshal Sokolov noted that "the military situation has become worse of late. Incidents of bases being shelled have gone up. . . . Such a war cannot be won through military means." Nikolai Ryzhkov, chairman of the Council of Ministers, noted that for the first time the information provided to the Politburo seemed to be objective.[86]

Yet even as the situation appeared to grow more difficult, key members of the Politburo dug in their heels. A major issue for Gorbachev remained the prestige of the Soviet Union in the Third World. Shevardnadze, perhaps influenced by Vladimir Kriuchkov, the chief of the First Directorate, who often traveled with him to Kabul, became Najib's biggest supporter at Politburo meetings. Indeed, in February 1987, it was Gromyko, a key player in the 1979 decision to invade, who urged a quick withdrawal, pointing out that "a half year more or less" of Soviet presence in that country would not make a difference. But Shevardnadze thought otherwise: "The most important thing is not to allow the Najibullah regime to fall. That is the most important thing!"[87]

Although Kriuchkov and Shevardnadze saw eye to eye on Afghanistan, they framed their arguments in slightly different ways. Kriuchkov's arguments were formed in more traditional geo-political terms. If the Soviet Union withdrew too quickly, Afghanistan would become a base for "Iran, Turkey, and fundamentalists." The Soviet Union could not just "leave, run, dropping everything. First we did it [invaded] without thinking, and now we will drop everything." Shevardnadze's logic was more subtle. What he saw in Afghanistan convinced him that the USSR had done so much damage in that country, there was almost no chance that a "friendly" Afghanistan could be preserved without a friendly leader. Although he opposed the war and supported the withdrawal, Shevardnadze did not see how Afghanistan would stay "friendly" or even neutral without a strong man to keep it that way. Najib needed to be trusted, since "there is not a family or village that has not suffered as a result of our presence. Anti-Sovietism will exist in Afghanistan for a long time," Shevardnadze told the Politburo. "Therefore, we need to have our own strong person in charge of Afghanistan."[88]

When National Reconciliation was first being conceived, it was assumed that the PDPA would still be the key force in the government. At the 20th PDPA Plenum, Najib assured his colleagues: "We will not retreat an inch from the achievements of the Saur Revolution. That is to say, in politics [those] who come to us should officially recognize the leading role of the PDPA and the people's power."[89] It was becoming clear, however, that even if Najibullah could somehow stay in power, the PDPA would not. Shevardnadze admitted that "the PDPA could collapse" at the next big turn of events.[90] Moscow began to look for ways that it could preserve a role for Najib without having to rely on the PDPA.

Some of the ideas that the Soviet leadership began discussing in the late spring and summer of 1987 seem to have been influenced by materials provided by academic specialists. One of these was Yurii Gankovskii, who had been trying to make his views heard by decision makers since 1980. At that time, Gankovskii had warned that unless Karmal was able to broaden the base of the regime, a civil war would quickly break out. His cautionary remarks fell on deaf ears.[91] In May 1987 he submitted a memorandum to Cherniaev, Gorbachev's foreign-policy aide, urging a more radical reformation of the Afghan government. It was a crucial moment, since Gorbachev and his colleagues were growing increasingly frustrated with officials in Kabul and their efforts to improve the situation there. Gankovskii argued that Soviet interests would best be served if the PDPA could be moved temporarily into the background. This would require a government head without a clear party affiliation, someone who was respected within Afghanistan as well as in other Muslim countries. This person could then reach out to the Islamic Conference and Muslim states, gaining international recognition. Only this kind of politician, argued Gankovskii, could even have a chance of making National Reconciliation successful.[92]

Since access to Soviet archives is currently limited, one cannot determine the extent to which Gankovskii or any other academics had an impact on policy making, or for that matter at which point policy makers really began to take scholarly views into account. Presumably the memorandum discussed above was not the only one Gankovskii wrote on the matter, and he may also have made his views known through informal meetings. Nevertheless, it is worth noting that in May 1987 key decision makers in the Politburo began thinking along the lines proposed in Gankovskii's memorandum. This would be consistent with the growing influence of other academics under Gorbachev.

Soviet leaders were becoming increasingly pessimistic regarding what they would be able to salvage from the USSR's position in Afghanistan. At the May 21, 1987, Politburo meeting, Gorbachev outlined what he thought a new regime might look like. Sectarianism was leading nowhere and would have to be eliminated. Although Gorbachev made it clear that he preferred to see Najib, rather than someone else, leading Afghanistan, he insisted that Najib should hold a state post, since he might then be able to portray himself as a truly national leader and have a chance of staying in power another year and a half. The Afghans would not follow Najib if he were a party leader, but "a president, a king, they would respect," Gorbachev said. Earlier discussions about opening 2–3 percent of government seats for Afghan émigrés were unrealistic; the proportion might have to be closer to 50 percent.[93] The emerging consensus in the Politburo was that the PDPA would be but one of the political forces in power after Soviet troops left. Even Kriuchkov and Gromyko agreed that reconciliation would mean accepting that the PDPA would lose its leading position.[94]

Soviet leaders hoped that some sort of new political stability could be achieved before their forces withdrew, while Moscow could still apply pressure both on its Afghan allies and on their enemies. Gorbachev was firm on the issue of power sharing in talks with Najib, telling the Afghan leader that the PDPA would have to give up government portfolios to opposition parties. The issue figured prominently during a July 1987 conversation. Gorbachev told Najibullah not to assume that the PDPA would stay in power but to begin inviting opposition figures into the government. This was the only way to face reality: "To count on the party's keeping its current position after reaching national reconciliation would be completely unrealistic." Gorbachev urged Najib to remain firm in the face of attacks from party members who were reluctant to share power—those who, in the spirit of "Karmalism," preferred loud slogans about the revolution, and were quite happy to have Soviet soldiers fight and die for them. Najibullah told Gorbachev he "agreed completely" and thought having the PDPA as a leading force was unrealistic in practice.[95] Similarly, Najibullah agreed that in November, at Gorbachev's insistence, he would nominate a prime minister from the opposition.[96]

In practice, Najib was reluctant to share power. He may have feared that if he alienated his party colleagues before securing some other source of authority, he would be left completely isolated. Kornienko, however, has written that this was because Najibullah offered only empty portfolios rather than important government positions. According to Kornienko, in the conversations between Najibullah and Shevardnadze, commitments previously made to Gorbachev were watered down with the foreign minister's consent, allowing Najibullah to avoid making any real movement toward power sharing.[97] It is unfortunate that there are no records available of those conversations so that historians could evaluate Kornienko's accusations. It is clear, however, that promises made to Gorbachev often did not lead to concrete results. At the time of the Soviet withdrawal in 1989, the government was still controlled by a shaky PDPA.

By the summer of 1987, it was becoming clear that National Reconciliation had failed to unite the party or to make the PDPA government more acceptable to the people of Afghanistan. Colonel Kim Tsagolov sent a long memorandum addressed to Dmitrii Iazov, the minister of defense, which touched on almost every major problem of the war, of governance, and of Soviet hopes and illusions in Afghanistan. Not only had the Policy of National Reconciliation failed to unite the PDPA, but it had completely failed to find any support among the opposition or even among other "democratic" parties. Tsagolov urged a radical change in course: "The PDPA is objectively moving toward its political death. No actions aimed at resuscitating the PDPA would produce any practical results. Najib's efforts in this respect can only prolong the death throes—they cannot save the PDPA from its death."[98]

Indeed, Gorbachev had largely given up on the idea of keeping the PDPA in power, and was starting to accept that the only government that could survive in the longer term was one which consisted in large part of opposition figures, albeit with Najib at its head. Yet it was becoming clear that the process of forming a new coalition government was going to take much longer than expected, in part because of the PDPA's reluctance.

Reports coming in from Afghanistan confirmed that National Reconciliation was failing: the armed opposition was not coming over to the government.[99]

Gorbachev brought up these issues when he met with Najib in July. He underlined that the PDPA was still failing to reach out beyond Kabul: "We have been receiving information that decisions being made in Kabul are arriving [at the provinces] much weakened." He urged Najibullah to become more proactive in including other parties in the government:

> It seems that in the second stage of National Reconciliation the question of creating a coalition government will come up, a bloc of leftist-democratic forces. You cannot refuse to cooperate with those who have a different point of view. You need to create real pluralism in the society and in government offices. The right tactic would probably be to emphasize that which unites these forces, and this will be the Policy of National Reconciliation: the ceasing of military activity.[100]

Gorbachev emerged disappointed from his July 1987 talks with Najib. For several months, the Politburo had been discussing why National Reconciliation had stalled. The talks with Najib, Gorbachev told the Politburo on July 23, "showed that Karmalism has put down deep roots. Everyone has started moving, but they are thinking first of all of themselves; even Karmal is raising his head. There could be a crisis in connection with this."[101] The Afghan problem occupied his mind during the summer holiday, and from time to time he sent his thoughts to Cherniaev: "We were pulled into Afghanistan, and now we don't know to get out. . . . It's awful when you have to defend Brezhnev's policies."[102]

Nevertheless, Moscow's policy on Afghanistan had shifted tremendously. The leadership in Kabul had been changed; advisers had been recalled; a new crop of people had been assigned to help achieve reconciliation. Most important, Soviet leaders were being stripped of illusions about what they could accomplish in Afghanistan before bringing the troops home. And while efforts within Afghanistan were disappointing, there was still some reason to hope that an honorable exit could be arranged. US-Soviet relations seemed to be improving. At the July 23, 1987, Politburo meeting, Gorbachev suggested that a three-party meeting of the United States, the USSR, and Afghanistan was necessary.[103] The effort to get the United States involved in an agreement would dominate Moscow's Afghan policy from the fall of 1987 until the signing of the Geneva Accords in April 1988.

From 1985 to 1987, Moscow's Afghan policy was defined by an effort to end the war without sustaining a defeat. As the previous chapter showed, Gorbachev was almost as concerned as his predecessors about the damage a hasty Soviet withdrawal might do to Soviet prestige, particularly among his Third World partners. Yet Gorbachev was also committed to ending the war, and for the most part had the support of his Politburo to do so. This meant looking for new approaches to developing a viable regime in Kabul that could outlast the presence of Soviet troops.

With regard to its policy in Afghanistan, Soviet officials continued to operate on the premise that the Afghan government could be made acceptable to the population with a combination of economic and political measures. Hence, Moscow invested much of its own money to help the Kabul government achieve legitimacy, and looked for ways to attract funds from its Eastern European satellites as well. Equally crucial were the political efforts: the replacement of Karmal with Najibullah, the launching of National Reconciliation, the efforts to broker a truce with certain rebels and the PDPA government. Much more so now than even during the Andropov era, these efforts showed a willingness to practice Realpolitik. Moscow was content to see a government that was Islamic in form, as long as it would remain friendly to the USSR. As with domestic reforms, however, decisions made by the Politburo were not implemented properly by officials in Kabul. The imperial attitude of Soviet advisers changed only slowly, while PDPA officials resisted efforts to curb their position and their privileges.

During this period, the Politburo received its first truly honest assessment of the situation in Afghanistan. At the January 21, 1987, Politburo meeting, Ryzhkov responded to Shevardnadze's report by saying that the Politburo was hearing a devastating account of the war "for the first time." Yet KGB chairman Chebrikov was equally correct when he said that such information had been available before. Previous chapters have shown that very skeptical and critical assessments had come from the military, as well as from other quarters, as early as 1980. What had changed was the Politburo's willingness to look at this information objectively, as well as to invite it into their discussions. Moreover, the questions were being discussed with the full participation of the Politburo—in contrast to procedures in the Brezhnev and Andropov years, when the Afghan Commission presented policies that were approved without much discussion.

While everyone in Moscow now recognized the apparent hopelessness of the situation in Afghanistan, they worried about the damage that a collapse there would have on Soviet interests. It became clear to Shevardnadze and others in the Politburo that, after seven years of war, the Afghan population was unlikely to think positively of the Soviet Union. This, in turn, meant that abandoning the PDPA completely was out of the question. Even as Soviet leaders lost hope in the spring of 1987 that a viable PDPA-led government could be constructed, they continued to look for ways to preserve a role for the party, or at least for Najibullah. Even as their faith in the party as a whole declined, their confidence in Najibullah grew.

The diplomatic efforts, which had been revived so quickly after Gorbachev came to power, were stalled for two reasons. One was the difficulty of talking to the United States about Afghanistan. US officials were unimpressed by the "signals" being sent by Moscow and continued to treat them as political ploys. In fact, it was only in the fall of 1987 that the United States began to take Moscow's desire to end the war seriously. Yet the Soviet-American relationship was only one part of a much larger problem. Another was that as long as Moscow held out hope that it could engineer a solution within Afghanistan, it would not separate its diplomatic initiatives from efforts within the country. Indeed, when the process seemed to be making significant progress in 1986, Moscow decided to link the issue of National Reconciliation to the withdrawal of troops. As long as Soviet leaders held out hope that National Reconciliation could work, they refused to consider uncoupling the two.

The summer of 1987 was a crucial turning point in the development of Gorbachev's thinking about reform, in the history of perestroika, and in the history of the Soviet Union. It was at this point that he told some of his closest advisers that he was prepared to change "the whole [Soviet] system, from economy to mentality."[104] Soon he would start speaking openly of de-Stalinization, a topic that had not been broached by Soviet leaders since Khrushchev. Although Gorbachev still spoke of putting pressure on Western countries in conversations with Third World leaders, he was increasingly eager to achieve a breakthrough in relations with the United States, even if Reagan did not abandon the hated Strategic Defense Initiative (SDI).[105] Not surprisingly, this was also a turning point in his thinking on Afghanistan and for Moscow's Afghan policy. It was becoming clear that National Reconciliation would not greatly increase the stability of the DRA government, that Najib was not a savior, and that the war could drag on endlessly. Gorbachev was losing confidence in Moscow's ability to change the situation in a meaningful way and to undo the errors of his predecessors with minimal political cost. It was at this point that he decided to turn to the United States directly. Though he still would have liked to see an Afghanistan in which a transformed PDPA played a key role, he now seemed ready to face the ultimate defeat of the regime after the withdrawal of troops.

5

ENGAGING WITH THE AMERICANS

The Geneva Accords, signed in April 1988, were the starting point of the Soviet withdrawal from Afghanistan. Since 1985, General Secretary Mikhail Gorbachev had been looking for ways to steer the Soviet Union out of the conflict without undermining Soviet prestige or leaving himself politically vulnerable. Successive efforts to shore up the communist government had failed, however: the surge of troops in 1985, the changing of the leadership in 1986, and the focus on National Reconciliation in 1987 produced only modest results. The Kabul government was still weak, and Soviet troops were still dying. Thus, at the end of 1987 the focus shifted from trying to change the military and political situation in Afghanistan to engaging in diplomacy with the other powers involved in the conflict, primarily the United States but also Pakistan. Although many of the details of the policy making remain murky, it is clear that Gorbachev was determined to get out. By April, it was also clear that the accords would be little more than a fig leaf for the withdrawal. Since both sides would continue to supply weapons to their clients, the conflict would continue, with the balance quite possibly falling against the government, which would no longer have the support of Soviet troops.

From the fall of 1987 to the spring of 1988, Gorbachev and his colleagues sought to use the improving US-Soviet relationship to secure an agreement on Afghanistan of the sort that had previously eluded him and his predecessors. Not coincidentally, this was a crucial period in Gorbachev's thinking about both domestic reform and foreign policy, and was likewise critical to the fate of the USSR. Since 1985, Gorbachev had followed a cautious approach to reform, often, as in the case of the antialcohol campaign, borrowing from Andropov's playbook.[1] In foreign policy, there were more genuine innovations; but as the failure of the Reykjavik summit showed, huge chasms remained in relations with the Reagan administration, and neither Gorbachev nor other Soviet officials showed any inclination to move away from Soviet commitments in the Third World. In the summer of 1987, Gorbachev told his advisers that he had come to see the need for more radical approaches to both domestic and foreign policy.[2] Having previously excluded "solidarity" with progressive regimes and movements from the range of topics that could be discussed bilaterally with the United States, he now decided to engage Reagan fully on the Afghanistan issue.

Determined to withdraw troops and improve relations with the West, Gorbachev was ultimately willing to sacrifice the long-standing Soviet position on stopping the supply of arms to the Afghan resistance. By the summer of 1987, it was clear that Soviet efforts to establish a viable regime in Kabul, including the ones undertaken since Gorbachev's rise to power, had failed. Yet in the fall of 1987, Gorbachev did not abandon hope of achieving a settlement in Afghanistan. Rather, he hoped that improving relations with the United States would lead to a settlement in Afghanistan and elsewhere in the Third World. In the end, Gorbachev's misjudgment of American politics and decision making and his inability to renege on traditional Soviet commitments meant that a Soviet withdrawal did not lead to a resolution of the conflict.

The US-Soviet Relationship and Afghanistan

Even before Gorbachev came to power, the worst of the "second Cold War" was over. Since 1983, Reagan had been moving away from the hard-right positions of his most conservative supporters. During the Able Archer NATO exercises that year, Reagan had been shocked to learn that Soviet leaders had nearly misinterpreted the maneuvers as the prelude to an actual attack. "The more experience I had with Soviet leaders and other heads of state who knew them," he recollected later, "the more I began to realize that many Soviet officials feared us not only as adversaries but as potential aggressors who might hurl nuclear weapons at them in a first strike."[3] Reagan went through a personal conversion of a sort: from that moment, his interest in engaging Soviet leaders, particularly on nuclear weapons, grew steadily.[4] Furthermore, moderates were on the ascendant in the Reagan White House and cabinet; Reagan was increasingly siding with those like US secretary of state George Shultz, national security adviser Robert "Bud" McFarlane, and the ambassador to the USSR, Jack Matlock, all of whom who favored negotiating with Moscow on a wide range of issues.[5]

By the time Gorbachev came to power, Reagan was desperate to engage with a Soviet leader. (They kept dying on him, he complained.) Although Reagan's primary concern was avoiding nuclear war, he was willing to discuss a number of issues, including Afghanistan.[6] But the early interactions between these two leaders showed that the distances they had to bridge were vast. And even as Reagan's thinking on nuclear war led him to seek greater arms control, it had barely evolved when it came to Third World issues, including Afghanistan. Gorbachev mirrored the US leader's posture: he, too, wanted to ease the tensions in the relationship and shared Reagan's aversion to nuclear weapons, but he did not trust Reagan enough to open a real bilateral dialogue on Afghanistan.

During their first meeting in Geneva, Reagan suggested a "coalition of Islamic states" that could supervise the installation of a new Afghan government. It is no surprise that this path, which would have left Moscow with little influence over the Kabul government, did not interest Gorbachev, and Reagan's follow-up letter was more conciliatory: "I want you to know that I am prepared to cooperate in any reasonable way to facilitate such withdrawal, and that I understand it must be done in a manner which does not damage Soviet security interests. During our meetings I mentioned one idea which I thought might be helpful, and I will welcome any further suggestions you may have."[7] In another letter, Reagan went further, telling Gorbachev that "withdrawal of [Soviet] forces" remained the only sticking point.[8] According to Jack Matlock—who, as a staff member on the National Security Council during Reagan's first term, had drafted the letter—the United States was prepared at this stage to stop aid to the *mujahadeen* if Soviet forces withdrew, without insisting that the Soviets cut off aid to the PDPA.[9] Gorbachev's wariness of the United States was still preventing the Soviets from engaging in direct negotiations and removing long-standing obstacles to the Geneva talks. Nor did Gorbachev believe that his relationship with Reagan had reached the point where they could profitably discuss regional issues. According to Cherniaev, in 1985 Gorbachev drafted guidelines on dealing with Reagan that included "not to get into regional issues; not to forgo our right to 'solidarity' with 'fighters for independence'; not to recognize US 'vital interests' indiscriminately, wherever it suits [the United States]."[10]

Nonetheless, US officials came away from the Reagan-Gorbachev Geneva Summit feeling that the Soviet attitude was indeed changing. The White House spokesman told journalist Don Oberdorfer that the United States had felt "something new" in Soviet policy, while the *New York Times* reported that "Mr. Reagan came away convinced that Mr. Gorbachev was looking for a diplomatic solution to the conflict in Afghanistan."[11] This did not yet translate into real bilateral discussions on the Afghan problem, however. In fact, the December 1985 round of the Geneva talks on Afghanistan proved to be the most frustrating of all. Despite renewed interest in diplomacy, neither the Afghan foreign minister, Mohammed Dost, nor his Soviet interlocutor, Nikolai Kozyrev, showed any flexibility. In February 1986, UN mediator Diego Cordovez traveled to Moscow, where he met with Shevardnadze and Georgii Kornienko. Shevardnadze seemed to promise "his help to break the deadlock." The round that took place after Karmal's replacement was more successful, with some progress on the time-frame issue. Still, US officials remained unimpressed. To them, the USSR did not seem genuinely interested in disengaging.[12]

By mid-1987, Soviet policy on Afghanistan had once again reached an impasse. Several successive strategies had failed to improve the stability of the Kabul regime, making it increasingly likely that an "honorable" withdrawal would be impossible. Although in November 1986 Gorbachev believed that the United States wanted to keep the Soviets in Afghanistan only to bleed the USSR, by the autumn of 1987 he was taking a new view of the US–Soviet relationship.[13] By autumn 1987, Shevardnadze and George Shultz had exchanged several useful visits, and a treaty on Intermediate-Range Nuclear Forces (INF) was nearly ready. It was logical that Gorbachev would try to use his improving relationship with the United States to achieve the settlement he found so elusive. The key point would be US willingness to stop supplying the opposition. Such a resolution would fully justify not only the Soviet withdrawal, but Gorbachev's entire foreign-policy framework, even to the most cautious and conservative elements in his own country and the communist bloc.

Gorbachev realized that he would first need to make it clear that the USSR was serious about withdrawal. Over the next six months, Gorbachev and Shevardnadze tried several times to use a tactic they had previously developed in negotiations with the United States: declaring an unexpected position as a starting place for negotiation. At the end of July 1987, Gorbachev told the Indonesian newspaper *Merdeka:* "In principle, Soviet troop withdrawal from Afghanistan has been decided upon. . . . We favor a short time frame for the withdrawal. However, interference in the internal affairs of Afghanistan must be stopped and its nonresumption guaranteed."[14] Soviet diplomats were told they could use the statement as a basis for saying the political decision had been made to withdraw. The statement was meant to jump-start negotiations and prompt the United States to agree to certain Soviet positions, making it clear that Soviet troops would withdraw in the hope that the Reagan administration would agree to earlier Soviet demands originally set as preconditions for withdrawal.

The first attempt to do this directly during a high-level meeting came during Shevardnadze's visit to Washington in September 1987. On September 16, Shevardnadze told George Shultz: "We will leave Afghanistan. It may be in five months or a year, but it is not a question of its happening in the remote future."[15] Shevardnadze asked the US secretary of state for cooperation in ensuring a "neutral, nonaligned Afghanistan." He also revealed that the Soviet leadership had made a firm decision on withdrawal.[16] In the context of the Geneva negotiations on Afghanistan, Shevardnadze's comment was a significant move, suggesting that the Soviet side would show its cards.[17] Similar statements had been made before, but this one convinced Shultz. The recent improvement in Soviet-US relations played an important role. As Shultz put it in his memoirs, part of the reason he accepted Shevardnadze's September 16 statement after rejecting earlier ones was that by then he "had enough confidence" to trust Shevardnadze's word.[18]

To what extent Soviet politicians favored this policy is unclear, but at least in its early stage it seems to have had support from Politburo members as well as senior Foreign Ministry officials. One measure of this, perhaps, is that similar feelers were put out on the eve of the Washington summit by KGB chairman Vladimir Kriuchkov, who would later adopt some of the most conservative positions on Afghanistan within the Soviet leadership. At a dinner meeting at the Maison Blanche bistro in Washington, he told his counterpart, CIA director Robert Gates, that the USSR wanted to get out but was seeking a political solution. Kriuchkov played on Gorbachev's theme of "mutual interests," emphasizing that a fundamentalist state in Afghanistan, if it took shape, would complicate US interests in the Persian Gulf. As Gates puts it in his memoir, Kriuchkov told him: "You seem fully occupied in trying to deal with just one fundamentalist state."[19] Other senior figures, both at the party level and at the deputy-ministerial level, also signaled that the Soviet Union was getting reading to withdraw. Early in November 1987, for example, Soviet Foreign Ministry spokesman Gennadi Gerasimov remarked that it would be possible for Soviet troops to leave within seven to twelve months. Toward the end of November, both Politburo member Nikolai Ryzhkov, speaking in New Delhi, and deputy foreign minister Igor Rogachev, speaking in Moscow, suggested that Moscow was ready to make an offer on the time frame.[20]

Gorbachev hoped that the improving US-Soviet relationship (or the improving Reagan-Gorbachev relationship) would make it possible to reach an acceptable agreement. He expressed this idea in a meeting with Najibullah in Moscow on November 3: "Maybe at the sunset of its rule the Reagan administration will want to show that it contributed—along with the USSR—to the settlement of the situation in a hot spot such as Afghanistan." For the moment, the US attitude remained unacceptable, because the Reagan administration "would want a settlement in which the PDPA would be pushed to the back"—but that could change. After all, the PDPA government represented the political and military reality.[21] Gorbachev believed that he could get Reagan to accept this status quo if Soviet troops withdrew.

The test of Gorbachev's new approach was the Washington summit in December 1987. Although the keystone of the summit was the INF treaty, for Gorbachev its importance lay not only in facilitating arms control talks but in being able to set the US-Soviet relationship solidly on a new footing. This included regional conflicts and, in particular, Afghanistan. Determined to explore the possibility that he could get a concession out of Reagan, Gorbachev pressed the issue during at least two meetings with Reagan and Shultz and one with vice president George H. W. Bush. The interpretation of these meetings greatly affected Soviet actions in the weeks that followed. Shevardnadze and Gorbachev seemed to believe that they had secured an important understanding regarding arms supplies, while the United States denied that any such concession had ever been made.[22] In fact, there was good reason for Soviet leaders to think that a concession had been made. At the same time, there was reason for them to be skeptical.

In the first conversation, on December 9, Reagan urged Gorbachev to move forward with an announcement regarding the start of the withdrawal. Although he promised that the United States would do everything to ensure that Afghanistan would become a neutral state, he balked at Gorbachev's request that the United States stop supplying the *mujahadeen*. Gorbachev had again tried to take the initiative by promising a quick end to Soviet participation in the RA's military operations: "I can tell you that the day the announcement is made about the withdrawal of Soviet troops, they will not participate in military operations, except for self-defense."[23] Reagan stuck to a familiar motif justifying continuation of US arms supplies: "The president of Afghanistan has an army; the opposition does not. Therefore we cannot ask one side to put down their arms while the other keeps them."[24]

The next day, Reagan's position seemed even less compromising. He suggested that the RA government should disband the army. Gorbachev insisted that there could be no question of troop withdrawal if the United States did not agree to stop supplying the opposition: "Only under the condition that it is tied with the question of stopping US aid to the opposition forces; that is, the day Soviet troops start withdrawing should be the day that American military aid is stopped."[25] If not, Gorbachev pointed out, the situation in the country would deteriorate, "making a Soviet withdrawal impossible." Here he tried the tactic he had earlier mentioned to Najibullah during their November meeting. Perhaps he could entice Reagan with the promise of a major diplomatic resolution. He suggested that it was time the United States and the USSR made a move together: "And regarding the cessation of American aid to the Afghan opposition: let's agree on a timetable and announce it. And if you need more time to think, then please do think. But we are inviting you to take a concrete joint step. This would allow us to see if the US administration is genuinely trying to find a solution to the situation in Afghanistan." Shultz, perhaps indeed thrown off balance by this last statement, remarked: "At the Geneva talks, a suggestion was made that the US could stop supplying Afghan freedom fighters deadly weapons sixty days after the start of the Soviet withdrawal."[26]

According to both Russian and US records, Gorbachev was firm on the point that the United States must stop supplying arms to the opposition. His position remained consistent with the brief prepared by the chief of the General Staff, Marshal Akhromeev. The brief stressed two basic preconditions: the end of arms supplies to the resistance and a guarantee of neutrality for any future Afghan government. Other issues, like the timetable for withdrawal, were more flexible. The troop withdrawal could easily be completed in less than twelve months, as long as other issues were settled.[27] Reagan proved largely unreceptive to Gorbachev's demands, insisting that if the United States cut off arms supplies, this would amount to an unacceptable "monopoly of force" for the Najibullah government.[28] Curiously, Shultz did seem to endorse the possibility of cutting off arms, remarking that the United States, like the USSR, supported the Geneva agreements, which stipulated that outside support to the opposition would cease sixty days after the start of the Soviet withdrawal.[29]

Thus, Shultz showed a willingness to meet the Soviets on the issue of arms supplies, and it seems that this had been considered by mid-level diplomats. Steve Coll, in his extensive study of the US involvement in Afghanistan, points out that US negotiators had been preparing to accept an end to CIA involvement around this time, while in late 1987 the American press treated the question of arms supplies as settled.[30] The confusion reflected the split between "bleeders" and "dealers" in the Reagan administration, as well as differences between State Department officials on the one hand and CIA officials and the vocal "Afghan lobby" in Congress on the other. ("Bleeders" preferred to see the Soviet Union remain in Afghanistan and take losses from the US-supplied *mujahadeen*. "Dealers" were those who believed it was better to compromise with Moscow, to encourage a quick withdrawal.) It is possible that Shultz was trying to maneuver Reagan toward his department's position, while Reagan was mindful of the political pressure he might face if he "abandoned" the *mujahadeen* and let them face the RA army alone. Although Reagan had said in a television interview prior to the summit that the United States would not stop sending arms, senior officials after the summit reaffirmed that the United States was in fact prepared to stop and that the main sticking point remained an acceptable timetable.[31] These confusing signals would have serious consequences for Gorbachev's view of his prospects for a suitable agreement.

Gorbachev apparently left the meeting believing that he and the US administration had reached a new understanding.[32] There was certainly reason for him to think that this was the case, although he should have remained suspicious. There had been no official agreement, nothing made public in the communiqué.[33] Gorbachev's belief that he was finally reaching a new understanding with the Reagan administration—an agreement that would lead to increased cooperation—defined Soviet policy in the weeks following the summit and affected the way the Geneva negotiations ultimately played out.

Initiatives and Concessions

While Gorbachev was negotiating with Reagan in Washington, one of the largest battles of the war was under way near Khost. It did not take long for the *mujahadeen* to recover their positions after the previous two operations in Zhawar, and by the fall of 1987 Haqqani's forces were blockading the road between Khost and Gardez. Inside Khost, 40,000 civilians and 8,000 soldiers had to be resupplied by air. Negotiations went nowhere—Haqqani was determined to take Khost and establish his beachhead. Toward the end of November, the Soviets and the RA army began a joint operation to relieve the garrison, led by General Boris Gromov on the Soviet side and the Afghan minister of defense, Shahnawaz Tanai, on the RA side. While the Afghans attacked the *mujahadeen* on the ground, moving from Gardez toward Khost, the Soviets pummeled them from the air. After another attempt at negotiations, the Soviets threw in their own paratroopers. On December 30, the first supply columns started moving along the roads.[34]

Operation Magistral, as it was called, was yet another qualified success. The Afghan troops had performed well, and the Soviets had demonstrated yet again that the *mujahadeen* could not stand up to them in conventional battle. Psychologically, this would have been a boost for the Najibullah regime. But soon after Soviet forces were pulled back in January of 1988, Haqqani's men returned to their positions. Khost was once again isolated, and had to be resupplied by air.

The effect of these events on Gorbachev's plans for withdrawal is unclear. Gorbachev had already made up his mind the previous summer. His main concern now was getting the most out of the Americans, in particular a US guarantee of noninterference. Such a guarantee would show that the United States was easing its demands as a positive response to the Soviet initiative. It would also change the dynamics of the fighting in Afghanistan by removing the element of "outside interference" and thereby justifying a Soviet withdrawal. The guarantee would provide a cushion for Najibullah once the Soviet withdrawal began. Within a highly factionalized government, Najibullah would not be able to hold power for long if it seemed he was being left alone against a US-backed opposition. Firm guarantees that US supplies would cease could strengthen Najibullah's position, perhaps even allowing him to achieve some of the goals set out in the Policy of National Reconciliation. Finally, Soviet leaders always had to contend with the possibility that their Afghan clients, although their geo-strategic influence was minimal, could act as spoilers, refusing to sign the accords if they felt their interests were not being addressed. This would undermine the possibility of withdrawal, create an unnecessary public breach between Moscow and an ally, and destroy the credibility of New Thinking in front of the world.

Shevardnadze traveled to Kabul on January 4, 1988, to talk with Najibullah and senior Afghan leaders. The main topic of discussion, of course, was the progress of negotiations.[35] Although the records of this conversation are not available, the timing, as well as the statements made by Shevardnadze before and after the trip, suggests that he felt some pressure to reaffirm a commitment to the Najibullah regime. In particular, he stressed that any agreement endorsed by the USSR and the United States would mean an end to arms supplies to the opposition when Soviet troops withdrew. In an interview before his departure, Shevardnadze told the Bakhtar news agency: "The American side has agreed to act as a guarantor and, accordingly, to end its assistance to armed groups that are engaged in military operations in Afghanistan against the people's regime." This point formed the basis of the agreement that he and Najibullah had reached during their talks. If the United States made a commitment to end outside interference, Soviet troops could begin their withdrawal and complete it in less than twelve months.[36]

While Gorbachev was increasingly coming to terms with the idea that Najibullah might not retain power, Shevardnadze believed that the USSR had a responsibility not only to work for a neutral Afghanistan, but also to help Najibullah stay in charge. General Liakhovskii, who was present at many of Shevardnadze's meetings in Kabul, believes that the "personal" factor played a big role in the foreign minister's relentless support of Najibullah.[37] During the January 6 interview with the Bakhtar news agency, Shevardnadze spoke of the need for the USSR to leave Afghanistan with a "clear conscience," which meant providing assurances that supplies to the opposition would end. Shevardnadze's sense of a "personal" commitment, combined with his belief that a "strong man" would be needed in Afghanistan, certainly played a role. Tellingly, in one of the few pages on Afghanistan in his memoirs, he wrote that he was bothered by a sense that the USSR was "abandoning" its Afghan friends, although he also noted that he had other worries besides his personal commitment to Najibullah.[38]

During the first few months of 1988, Shevardnadze still hoped that the Geneva Accords could become a proper instrument of guarantees and enforcement. That way, they could give Najibullah a chance of surviving and protect Soviet credibility with other Third World countries. On January 15, 1988, he told his Politburo colleagues that National Reconciliation was having an effect and that the PDPA would be able to play a leading role in the government if it could avoid factionalism.[39] By contrast, in May of the previous year he had reported that the effect of National Reconciliation had been quite limited. This new line reflected Shevardnadze's growing faith in Najibullah, as well as his belief that without a strong pro-Soviet leader Afghanistan would not remain a friendly country.

While the idea of ending supplies to the *mujahadeen* in exchange for a Soviet withdrawal may have been acceptable to US negotiators, it proved unpalatable to the Reagan administration, perhaps because it was so politically risky. Shevardnadze's interview on January 6 supposedly surprised and angered US secretary of state George Shultz, who immediately sent a telegram to Moscow stating that Reagan had never made any such promise, and denied it publicly as well.[40] The incident put Moscow in a bind. It had promised Najibullah that the USSR could get the United States to stop supplying the opposition. Shevardnadze had followed this up publicly with an interview carried around the world, and had then been rebuffed by Shultz and Reagan in an equally public manner.

Shevardnadze also seems to have genuinely believed that the Geneva Accords could be more than just a fig leaf for the Soviet withdrawal. He insisted to subordinates that by signing the accords, Pakistan was binding itself to stop interference and would have to respect that agreement.[41] In a meeting with UN mediator Diego Cordovez in January 1988, he pressed for a strong enforcement mechanism so that the USSR could be reassured that "Pakistan would respect all the provisions of the agreement."[42] Pakistan's willingness to be bound by the accords was important not only for Afghanistan, but also for Shevardnadze's relative standing within the Soviet leadership.

Reagan's public reversal on supplying the opposition threatened to undermine Gorbachev's and Shevardnadze's positions with the "conservative" elements in the military and the government.[43] While most segments of Soviet bureaucracy were in favor of withdrawal, there were still differences over the manner in which the withdrawal should take place and in which Moscow's relationship with Kabul should evolve. The military, for example, favored either a unilateral withdrawal or one conducted through the Geneva process, but only if it provided concrete guarantees of parallel disengagement on the part of Pakistan. General Varennikov wrote that his team petitioned Moscow numerous times to work for symmetry in withdrawal. He suggested to both Shevardnadze and Cordovez that for every military facility Soviet troops left, Pakistan should dismantle one of the *mujahadeen* facilities on its territory.[44] According to Liakhovskii, the top Soviet military leadership in Afghanistan felt that the Geneva process was pointless unless it brought real guarantees of the kind Varennikov demanded.[45] Aside from trying to lobby Shevardnadze, Gorbachev, and the Politburo, however, they could do little in terms of affecting the Geneva process. Reagan's flat refusal to provide such guarantees made Gorbachev's recent enthusiasm for an agreement with the United States seem foolish, and could have become fodder for conservative critics if withdrawal was followed by disaster in Kabul or if the still-nascent rapprochement in US-Soviet relations collapsed.

With Reagan and Shultz rejecting the possibility that they would cease supplying the *mujahadeen* in exchange for a Soviet withdrawal, Gorbachev and the Politburo were faced with a stark choice. They could either retrench, refusing any further concessions until the United States agreed to stop weapons supplies, or push forward, hoping that the United States would come around if conservative "bleeders"—who, Moscow believed (correctly), were responsible for America's hard-line policy—could be convinced that the intent to withdraw troops was genuine. The first approach was the one favored by the military, by Shevardnadze, and to some degree by the Foreign Ministry negotiators in Geneva. The danger, however, was that such a retrenchment could stall the whole withdrawal process, leaving Soviet troops in Afghanistan because of diplomatic hurdles.

At this critical moment—with the talks stalled, the Reagan administration proving completely uncooperative, and Soviet officials and advisers at loggerheads about how to proceed—Gorbachev opted for yet another bold, unilateral announcement.[46] In a statement aired on Soviet television, he announced that the Soviet withdrawal would begin on May 15. Commitment to a start date for the withdrawal had been a long-standing American demand, and Gorbachev was hoping that by committing to a date he could nudge the Americans to revisit the issue of arms supplies. Georgii Kornienko, the deputy foreign minister, claims that he introduced the idea of the announcement in the belief that such a statement from Gorbachev would accelerate the Geneva process.[47] Bolstered by comments made by US officials during his trip to Washington in January 1988, Kornienko argued that announcing a withdrawal date would allow the United States to apply greater pressure on Pakistan and would convince Najibullah to sign. Shevardnadze rejected this approach, agreeing only to a statement to the effect that "a withdrawal of troops could begin in May 1988 if a settlement agreement could be signed in February–March." The Politburo accepted this phrasing, and Shevardnadze and Kriuchkov conveyed the message when they went to see Najibullah in Kabul to discuss the planned announcement. At the last minute, however, Gorbachev opted for stronger wording, personally writing it into the Politburo decision by hand.[48]

Kornienko claims he played a key role in this last-minute decision, but it is quite consistent with Gorbachev's preferences in similar situations. An announced start date had been a frequent demand of the Reagan administration, repeated during the Washington summit and frequently in the press. Gorbachev chose to make the announcement in order to take the initiative—to do what his counterparts in the United States doubted he would do. The disagreement between him and Shevardnadze was the sort that can arise between a politician and a negotiator. Although Shevardnadze was often more the former than the latter, in this instance he saw that the announcement meant going into the next round of negotiations holding fewer cards than ever. This, too, was consistent with the role that Shevardnadze had been playing in the previous months: that of Najibullah's top ally in Moscow and chief negotiator on the international scene.

The statement, read on Soviet television on February 8 and printed in both *Pravda* and *Izvestia*, committed the USSR to start the withdrawal on May 15 as long as an agreement had been reached at Geneva by March 15. It also committed the USSR to "front-load" the withdrawal—that is, to include a larger proportion of troops in the first half of the withdrawal. Front-loading had been a US and Pakistani demand, intended to make sure any partial withdrawal was irreversible. Finally, the statement made it clear that the withdrawal would be de-linked from the formation of a coalition government, an earlier Pakistani demand.[49] There was only a single mention of noninterference, as "one of the aspects of the settlement." Najibullah released a parallel statement the same day. It is unfortunate that the records of Kriuchkov and Shevardnadze's meetings with Najibullah that week are unavailable, as they would make for lively reading. It is highly unlikely that Najibullah was particularly enthusiastic.

With regard to the talks in Geneva, the February 8 announcement, made despite Shevardnadze's opposition, had the desired effect. The Pakistani president, Zia al-Huq, previously noncommittal regarding Pakistan's role in the last stage of negotiations, now told Cordovez that Pakistan would "fully participate" in the upcoming talks.[50] Although there would be further hurdles prior to signing the Geneva agreement, February 8, 1988, became a turning point. Cordovez describes it as the breakthrough he had been waiting for, allowing him to announce that the talks would resume on March 2.[51]

While the withdrawal announcement facilitated the Geneva process, it undermined the Soviet position at the talks. From the point of view of Soviet negotiators, any flexibility on their part was met with a firmer hand from Islamabad and Washington.[52] In his analysis of the accords, Soviet negotiator Nikolai Kozyrev pointed out that prior to December 1987, statements regarding the Geneva process made in Moscow had reflected recommendations made by the Soviet team in Geneva. After December, the statements were often made without consulting or warning the Geneva team.[53] Shevardnadze's staff in Geneva had opposed previous announcements, such as the one made by Gorbachev in December 1987 and the one made by Shevardnadze in January 1988, that the USSR would be willing to withdraw its troops within twelve months in exchange for the creation of a broad coalition government in Kabul and the cessation of aid to the *mujahadeen*. According to Kozyrev, these announcements "devalued the position of our delegation at the talks, put it in an awkward spot, and gave the opposite side extra motivation to pressure Moscow in the hope that the Soviet leadership would agree to further concessions."[54] Even as they pushed the talks toward an agreement, Gorbachev's unexpected announcements took away some of the leverage that Soviet diplomats hoped to employ in negotiations.

The withdrawal announcement should be seen in the context of Gorbachev's political style, as well as in the light of his changing conception of foreign policy in early 1988. While the comment on interference barely took up a line in the statement, some twelve sentences dealt with the connection between a resolution to the Afghan conflict and tensions in other Third World hot spots. This included the Iran-Iraq war and conflicts in southern Africa, Cambodia, and Central America. Using language he had previously employed to describe the Afghan war to the CPSU, Gorbachev called these hot spots "bleeding wounds capable of causing patches of gangrene on the body of mankind."[55] But if Gorbachev was a true believer in his reforms and in his vision of a new foreign policy focused on cooperation, as both his detractors and supporters say, then the linkage made sense. A politician's intuition told him that he was not the only leader dealing with a thorny problem. Reagan could be persuaded to see the mutual advantage of a new approach, but he would have to start in Afghanistan.

The February 8 statement was not pure propaganda. Several weeks later, meeting with Politburo members to hear a report on the Afghan situation, Aleksandr Iakovlev, a close Gorbachev aide and a Politburo member in charge of ideology (as well as a member of the Afghanistan Commission), told his colleagues to take this line as policy. The formal statement had been about Afghanistan, but "our announcement is a real solution for one regional conflict and a possible formula for others. Let us bring the same sense of responsibility and international participation to other regional problems, whether Angola and the SAR, or the Near East, or Central America."[56] For Gorbachev, the new formula was more important than losses at the negotiating table.

Gorbachev's announcement helped to clear one of the last hurdles to completing the formal Geneva document. The Pakistani side had demanded that a coalition government largely excluding the PDPA be formed prior to the withdrawal of Soviet troops.[57] Gorbachev's rejection of this proposal met with no resistance from Shultz, who had not found it reasonable and only reluctantly agreed to carry Zia's demand to Moscow.[58] As one senior official told reporters just after Shultz's Moscow trip, it would be wrong if the United States now asked the USSR to "stick around" until a political settlement had been reached.[59]

While the announcement furthered the negotiations at the Geneva level, the United States did not agree to stop supplying arms to the opposition. It had become clear to the US administration that the Soviet Union was desperate to leave, and there was no reason to take a political risk domestically by making any concessions.[60] Shevardnadze kept trying to convince the United States to agree to halt arms supplies with the start of the Soviet withdrawal. When Shultz came to Moscow in February, Shevardnadze accused the United States of "switching signals" on the question of arms supplies.[61] After all, the USSR had done all it was asked to do, including announcing a start date and offering a short timetable for the withdrawal. Shevardnadze emphasized that Najibullah was working toward a coalition government that would include the opposition while marginalizing the most extreme elements. Shultz remained adamant that a US cutoff would come only if the USSR also stopped supplying Kabul. In Washington the following month, Shevardnadze again pressed this point, but Shultz refused to back down. After the February meeting, Shultz said he "had no doubt that Soviet troops would be withdrawn."[62] Shultz knew that further concessions were unnecessary, that Moscow now wanted the accords more than the United States did. There was no need to take steps that would cause a conservative backlash back home.

Gorbachev should perhaps have realized that the Reagan administration would avoid a politically risky step if it could. Conservative commentators and politicians in the United States had been edgy since the Washington summit, worrying that Reagan would give away too much.[63] Even the mainstream press did not see any reason for stopping aid to the *mujahadeen*. The day after the February 8 announcement, the *Washington Post* argued that support for the *mujahadeen* was a duty of the United States, a responsibility "to sustain a brave people fighting to repel foreign aggression."[64] Gorbachev had easy access to this kind of information via the KGB and the Foreign Ministry, and should have realized that Reagan had nothing to gain politically by stopping supplies to the Afghan opposition.

Despite having lost on the key issue of arms supplies, Gorbachev held out hope that a new, broader understanding with the United States would lead to a peaceful resolution of the conflict sometime after Soviet troops withdrew. Shultz did accept Gorbachev's broader framework for conflict resolution. In a closing meeting on February 22, he pointed out that the most valuable parts of the visit were understandings about how conflicts in Angola, Cambodia, and Iran-Iraq could be settled. Gorbachev concurred: "I think we have to set an example for the world in these questions. If we develop this sort of cooperation, one can hope that conflicts will be decided in a way that addresses the interests of all sides." The most curious thing about this conversation, however, was that the issue of a US aid cutoff was not even mentioned. Gorbachev restricted himself to urging Shultz to ensure that the next round of Geneva talks would be the last, and to underscoring that the USSR would not now accept a linkage between a coalition government and troop withdrawal. To this latter point, Shultz readily agreed.[65]

Toward the Geneva Accords

By mid-February 1988, Gorbachev had reconciled himself to the idea that a Soviet withdrawal would not bring about the cessation of US aid to the *mujahadeen*. Although Gorbachev was prepared to accept a weak agreement as long as it paved the way for Soviet troops to withdraw, Shevardnadze kept trying to push for a new agreement.

It is quite possible, in fact, that Gorbachev had made the February 8 announcement fully expecting to begin the withdrawal without a US agreement to cut off aid. On February 11, he seemed to be preparing the Indian minister of defense, Krishna Pant, for a Soviet acceptance of a weak accord. When Pant pointed out that US weapons could fall into the hands of rogue terrorists, Gorbachev replied that the question of arms supplies was difficult; but if the USSR pursued it, the United States could counter by pointing to Soviet weapons held by the Kabul regime, "and then the whole process could get stuck. And we don't want to leave Najib naked."[66] At the February 22 meeting with Shultz, Gorbachev did not bring up the question of arms supplies at all, suggesting that he was prepared to accept an agreement that did not stop the United States from supplying the opposition via Pakistan. He needed a withdrawal to prove he was serious about putting the Soviet Union on a new foreign-policy course. Agreement or not, the USSR had to withdraw. As he explained to his Politburo colleagues on March 3, "The country, the world, is ready for us to do this. In politics, it is not only what you do that matters, but also when and how."[67]

Shevardnadze could not accept such a stance. His close ties with Najibullah, developed over several years and numerous meetings, pushed him to seek an agreement that would help guarantee the regime's survival after Gorbachev had given up on this. During a trip to Washington in March, Shevardnadze made one final attempt to get the Reagan administration to stop supplying the *mujahadeen*. In a meeting with Shultz, he emphasized that Moscow had met all of Washington's earlier demands. The timetable had now been reduced to nine months and could be made even shorter, while the withdrawal would be front-loaded, meaning that half the Soviet troops would leave in the first ninety days. Shouldn't the United States respond by meeting one of Moscow's demands? Shultz rejected these arguments. The next day, Shevardnadze tried again. Shultz consulted with the US national security adviser, Lieutenant General Colin Powell, and with acting secretary of state Michael Armacost, and came back to tell Shevardnadze, again, that the United States would cut off aid only if the USSR did as well.[68] A few days later, however, they conferred over the telephone and agreed to set aside the question of cutting off arms supplies. Shultz confirmed this with a letter, and the stage was set for the accords to be signed.[69]

Throughout this period and after the troop withdrawal had begun, Shevardnadze and Kriuchkov formed a sort of "Najib" lobby within the Soviet leadership. Shevardnadze's trips to Afghanistan had convinced him that unless a strong leader was in charge, the country would become firmly anti-Soviet. It is unfortunate that no records are available of Shevardnadze's conversations with Najibullah, since these would reveal much about the dynamics of their relationship. Nikolai Egorychev, Moscow's ambassador to Kabul, has said that Shevardnadze guarded the relationship rather jealously.[70] According to Liakhovskii, Shevardnadze made extensive promises to Najibullah about Soviet support during their meetings.[71] Even in April, when the accords were about to be signed, Shevardnadze argued for a revision of the 1978 Soviet-Afghan Friendship Treaty to permit the return of Soviet troops under certain circumstances, but Gorbachev refused this approach.[72]

Gorbachev, however, viewed the withdrawal from Afghanistan as part of his overall political reforms, as well as crucial to the USSR's standing with its allies. As Iakovlev pointed out, the USSR absolutely had to get out; most Soviet people knew this and supported the decision. But Moscow had to keep the "national interest in mind." It was a question of authority and legitimacy: "After all, we have to explain this problem to all our people, to the mothers, to public opinion. We have to look at what the reaction will be like abroad. Some people will be unhappy with this step. We have to look really carefully at the reaction in the Third World."[73] Although achieving the broader goals of the New Political Thinking required ending the war in Afghanistan, it also meant preserving a sense of the USSR's power and authority, without which Moscow would very quickly lose its position in the world.

Gorbachev's public and private statements suggest that he would have preferred an agreement that preserved a neutral Afghanistan ruled by a broad coalition government incorporating the PDPA. Notwithstanding his commitment to the New Political Thinking, he remained concerned with the USSR's great-power status. He acknowledged that a withdrawal from Afghanistan that did not guarantee Najibullah's survival in power would invite challenges from conservatives within the USSR, as well as from socialist governments in Eastern Europe and the Third World. Nevertheless, Gorbachev told his Politburo colleagues on March 3, 1988, that challenges from the Third World and from conservatives should in no way affect the withdrawal decision: "There will be questions, even in our country. What did we fight for? What did we sacrifice so many for? In the Third World there will be questions. They're already coming in. You can't depend on the Soviet Union, they say. It leaves its friends to the mercy of the United States. And here we must not budge."[74] His commitment to withdrawal from Afghanistan was now absolute. He knew, however, that he did not operate in a political vacuum, that his foreign and domestic policies would invite criticism and opposition, and that he had to proceed carefully at every step. It was important, he pointed out, "to keep the authority of power before our own people and the outside world."[75]

The Politburo gathered in the late afternoon on April 1 to decide whether or not to sign the Geneva Accords. Gorbachev knew the accords were weak, but their existence gave him hope that the Soviets could affect how the withdrawal was to be played out. It is clear that he did not expect much from the accords themselves. Rather, he saw them as a symbol of the way he wanted to conduct relations, perhaps even a stepping stone that would help to establish trust. As he put it, "This will be a confirmation of our entire approach to solving international problems."[76] Every single member of the Politburo voted in favor.[77]

Najibullah knew that the withdrawal was inevitable. As he told an interviewer in 1989, he had taken Gorbachev seriously when the latter first came to power and began talking about Soviet disengagement.[78] Yet he also knew it would be incredibly difficult for his government to survive without the support of Soviet troops. Shevardnadze had clearly been promising him that the accords would not be signed unless the United States also agreed to stop supplying arms. Over the previous month, this position had disintegrated, and now Soviet troops were preparing to withdraw and leave the Afghan army, such as it was, fighting largely alone.[79] Shevardnadze did what he could to ensure continued Soviet support.

Najibullah's resistance to an agreement that would potentially weaken him was on full display when Shevardnadze flew to Kabul on April 3. According to deputy minister Yulii Vorontsov, who was involved in the Geneva process and later served as ambassador to Afghanistan, Najibullah at first refused to sign. It took Shevardnadze three days of difficult persuasion to make the Afghan leader agree to the accords.[80] It also took extensive promises of Soviet support and even the possibility of leaving 10,000–15,000 troops in the country.[81] Shevardnadze stayed in Kabul until April 5, then returned to Moscow and announced that Najibullah had accepted the agreement. The next day he flew to Tashkent with Gorbachev, Kriuchkov, and Cherniaev to meet with Najibullah. By the time Gorbachev met Najibullah in Tashkent on April 7, all of these questions had largely been solved, and Gorbachev assured Najibullah that the Soviet government completely endorsed the agreement reached between Shevardnadze and the Afghan leader over the previous days.[82]

The accords were a threat to Najibullah not only because they deprived him of Soviet troops without any cutoff of supplies to the *mujahadeen*, but also because his authority within the government could be further eroded as a consequence. This was highlighted when RA foreign minister Abdul Wakil refused to sign the accords in Geneva, saying that to do so would be to betray his people.[83] While Shevardnadze and Gorbachev met with Najibullah, Nikolai Kozyrev worked on Wakil in Geneva.[84] In his hotel room, Wakil put on a great show of emotion, ripping napkins and screaming that the Afghan people would never forgive him. Although Najibullah had agreed to the accords, Wakil still refused to sign, and relented only after Soviet deputy foreign minister Vorontsov, who flew to Geneva at Kozyrev's request and spent six hours alone with Wakil, made it clear that if the foreign minister did not sign, another official would be sent from Kabul.[85] The intense effort put in by Shevardnadze and his aides could only have increased the foreign minister's sense of commitment to the Najibullah regime. More important, it highlighted how fragile Najibullah's position could be if it was not absolutely clear that he had complete Soviet support.

Gorbachev needed the meeting in Tashkent just as much as Najibullah did. As he explained to Alessandro Natta, the General Secretary of the Italian Communist Party, the imminent withdrawal from Afghanistan was already causing rumblings among Soviet allies, particularly in the Third World. The essence of this criticism, according to Gorbachev, was: "You're 'abandoning' Afghanistan, and you will 'abandon' us."[86] While Gorbachev needed to demonstrate that the USSR really wanted to do business in a new way, he also needed to show that it was not about to leave its friends in the lurch. Proving that the Soviet Union could do both meant expressing confidence in Najibullah as a leader who could survive without the aid of Soviet troops. This was a key purpose of the April 7 meeting in Tashkent, a "heads-of-state meeting" that was supposed to represent the beginning of a new relationship between two sovereign states.

Gorbachev used the meeting to provide political cover for his approach. Between April 1, when the decision to sign the accords was discussed, and April 14, when the signing ceremony took place, Gorbachev personally briefed communist-bloc leaders and party bosses, telling them that Najibullah was a capable leader who was gaining in authority and that the USSR would continue to support him politically. This, and the usefulness of the Geneva Accords, were central themes of his conversations with Cuban leader Fidel Castro on April 5 and with Czechoslovak president Gustáv Husák on April 8.[87] At meetings with regional party secretaries called to Moscow in April after the Nina Andreeva affair, which had alerted Gorbachev to the strength of conservative feeling in the country and within the leadership, he again stressed the importance of withdrawal.[88] He admitted that there could be an unfavorable turn of events, but insisted the Geneva Accords would help settle the political crisis.[89]

Although Shevardnadze formed a united front with Gorbachev just prior to the signing of the accords, he was clearly unhappy with the result. On April 1, 1988, Shevardnadze told the Politburo that with the Geneva Accords there was a "legal basis" for the withdrawal, which meant that the United States could no longer use Pakistani bases to resupply the *mujahadeen*, and that there would be 150 monitors to make sure the accords were carried out.[90] In fact, Shevardnadze's support for the accords was half-hearted at best. He had fought hard to secure an agreement to end arms supplies. In his memoirs, Shevardnadze confesses that he left Geneva with mixed feelings: "I knew that we would not lessen our political efforts for a peaceful settlement in Afghanistan, but still I could not rid myself of a sense of personal guilt toward my friends."[91] The accords were a much weaker document than what many in the Soviet government and the PDPA had sought. Although they did contain noninterference clauses, the question of arms supplies was left open. The accords contained no guarantee of a role for the PDPA in a future government, and had only a weak enforcement mechanism. The latter point, in particular, greatly irked both the diplomats and the Soviet military.[92]

For Gorbachev, the Geneva Accords served a dual purpose: they could be used as a shield against conservatives as the withdrawal got under way and would serve, in the eyes of Westerners, as proof of the USSR's commitment to political solutions. Despite the weakness of the proposed document, Gorbachev argued that it was the best way to get out, in part because it would allow Moscow to maintain a degree of leverage in future discussions. The biggest caveat was the political victory. Arguing for signing the accords rather than for pursuing a unilateral withdrawal, he put the issue in the wider context of his domestic and international challenges: "It is hard to overestimate the political value of settling the Afghan problem. This will be a confirmation of our new approach to solving international problems. Our enemies and opponents will have their strongest arguments knocked out of their hands."[93] The brief discussion of Afghanistan in his memoir stresses this aspect: "The significance of this unprecedented settlement went far beyond its regional implications. It was the first time that the Soviet Union and the United States, together with the conflicting parties, had signed an agreement which paved the way for a political solution of the conflict."[94] Gorbachev was less concerned with the fate of Afghanistan than with the success of his broader foreign policy.

For all the rhetoric about changing the way conflicts were solved and the way the USSR behaved in its foreign relations, Gorbachev had to go to great lengths to show that in many ways things were still the same. One phrase in particular from the April 7 Tashkent meeting captures this. Seeking to reassure Najibullah that the USSR intended to keep supporting the regime with arms, Gorbachev framed the commitment in thoroughly uncompromising terms: "Even in the harshest, most difficult circumstances, even under conditions of strict control—in any situation, we will provide you with arms."[95] This was a far cry from the talk of mutual settlement of conflicts that had come from Gorbachev so often on previous occasions. As he often did throughout his tenure, Gorbachev maneuvered between two positions: one imaginative and reformist, the other much closer to traditional Soviet policy and priorities. His willingness to make contradictory promises and statements (albeit to different audiences) would come to haunt him during the withdrawal period, when some of his advisers expected him to stick to the letter of the accords, while others insisted that he honor his promises to Najibullah.

It was noted earlier that in mid-1987, Gorbachev—increasingly frustrated with the slow pace of reforms—began to see all of his country's problems as interlinked, and solvable either all at once or not at all.[96] This observation helps to explain his Afghanistan policy in the period discussed in this chapter. In previous periods, Gorbachev had approved various policies that would improve the situation within Afghanistan. Now he took a more direct, personal role to try to bring the Soviet intervention to an end. There were two reasons Gorbachev was willing to abandon seemingly strong negotiating positions. First, he hoped that this would help to achieve a broader improvement in relations with the United States. Second, and most important, he did not want to drag out the Soviet involvement in Afghanistan any longer, because by the end of 1987 he had lost faith in most Soviet military and political efforts within that country. Not coincidentally, this was also the period in which his thinking on the Third World began to change: no longer was competition with the West a crucial element in the way he thought about the countries which only months before he had talked about defending from imperialism. The main focus now was on conflict resolution. The situation in Afghanistan contributed to this change in thinking; at the same time, Gorbachev's approach to that problem was affected by his broader frustrations with Moscow's allies.[97] Gorbachev's initiative proved crucial in ensuring that a withdrawal date was announced in February 1988 and that the accords were signed in March. He overrode objections from the military, as well as from people closer to him, like Shevardnadze, in order to bring this about. Gorbachev chose this course because he believed that he could sacrifice a favorable settlement on Afghanistan for a broader improvement in relations with the West.

Many of the people around Gorbachev were not so sanguine about the chances that a US-Soviet rapprochement would lead to a favorable resolution in Afghanistan. Quite correctly, they saw that the Reagan administration was unwilling to give up supplying aid to the *mujahadeen;* not only did Reagan himself believe in the moral value of that aid, but there was a vocal congressional lobby that was skeptical of even his very limited engagement with the USSR on this and other regional issues.[98] More important, some of the most senior officials around Gorbachev were very closely involved with the Kabul leadership and saw themselves as responsible for representing the PDPA's interests. This included the KGB chairman Vladimir Kriuchkov and foreign minister Eduard Shevardnadze. Shevardnadze may indeed have felt a great deal of "personal" responsibility, but with Kriuchkov it was also an issue of maintaining Soviet (and KGB) commitments to client governments. Perhaps they saw the abandonment of Kabul as a precedent for the Soviet government to abandon all of its commitments—a domino effect starting from the center. Najibullah exploited this situation fully, securing promises from Shevardnadze and Kriuchkov in return for his cooperation in Gorbachev's diplomatic game. This explains the frequent references to protecting "friends"—a concern which Gorbachev acknowledged but was willing to set aside.

For all of his emphasis on the New Political Thinking, which was genuine, Gorbachev could not ignore such concerns. He had to think about his political strength at home and also about the USSR's relationship with its allies. His promise of support and arms supplies to Najibullah "no matter what" was only one example of his willingness to conduct relations with Third World client states much the way his predecessors had. On April 1, 1988, the day the Geneva Accords were discussed, he approved a major airlift of arms to Colonel Haile Miriam Mengistu's regime in Ethiopia—ignoring objections from Cherniaev, as well as reports from Marshal Akhromeev, the chief of staff, which showed that the situation was hopeless.[99] Similarly, for all the talk about applying New Political Thinking to other international problems, Gorbachev made no serious effort to incorporate Soviet aid to the Sandinistas or to the MPLA in Angola.

The "conservative" (in this case) critics who were unhappy with the accords did not offer viable alternatives, however, and this explains why support in the Politburo for signing the accords was nearly unanimous in March 1988. Delaying the withdrawal would have caused more Soviet deaths at a time when Gorbachev had already called the war a "bleeding wound." It might also have undercut the enormous leap he was about to take in US-Soviet relations. A unilateral withdrawal would have provided the USSR with greater freedom of action in the future, but it would have done the same for both the Pakistanis and the Americans. The Geneva Accords at least created a precedent for international agreement, and, by convincing all parties that the USSR was serious about pulling out, helped to achieve a relatively bloodless withdrawal.

In developing the New Political Thinking, Gorbachev had to reconcile two often-contradictory positions: maintaining Soviet prestige while increasing cooperation with the West. This was most difficult in relation to Afghanistan, where the minimum necessary to enable an "honorable" Soviet withdrawal was far from what was sought by the United States and its allies. The key issue, as Gorbachev saw it, was to build up trust, rather than continue to undermine it by stalling at negotiations, which in any case would lead to a prolonged stay in Afghanistan. It was no longer a question of "winning" in Afghanistan, but rather a question of converting the withdrawal into a foreign-policy triumph in other areas. As he put it to the Politburo after the accords were signed, "Having lost in Afghanistan, we have to win in the world."[100]

In April 1988, Gorbachev believed that the concessions made to the United States over the previous several months were worth the price, since they would lead to a new relationship between the two countries and the solution of problems in Third World hot spots. The behavior of Shevardnadze, Kriuchkov, and even Gorbachev showed that there was a limit to how far Moscow would go in backing away from supporting its client in Kabul.

6

THE ARMY WITHDRAWS AND THE POLITBURO DEBATES

In a memorandum written for Gorbachev in 1986, Aleksandr Iakovlev, the "architect of perestroika," argued that Soviet foreign-policy making was hampered by the competition and infighting that took place between various bodies, including the KGB and the military. The only way to overcome this chaos in Soviet decision making, Iakovlev said, was to create an agency to oversee the coordination of foreign-policy making, on the model of the US National Security Council.[1] Nowhere was Iakovlev's critique of the way Moscow made foreign policy more applicable than in the case of Afghanistan, where the military, the KGB, the party, and the Foreign Ministry adhered to different, often contradictory policies. All of them operated with the same ostensible goal in mind: to create a stable government in Afghanistan that could stand on its own two feet even after Soviet troops withdrew. In practice, each had its own view of how this was to be accomplished. Whereas the KGB believed that Kabul's best chance was a strong Najibullah who had the resources to establish his authority and become a Pashtun leader in the mold of Amir Abdur Rahman, the military favored dealing with Ahmad Shah Massoud, the Tajik commander who controlled opposition forces in the north of the country.

The signing of the Geneva Accords and the start of the troop withdrawal only exacerbated these divisions, which reflected not only disagreement over Soviet priorities in Afghanistan, but also very different assessments of the situation. To take one example, senior Soviet military officers in Afghanistan, organizing the transfer of over 100,000 troops and assorted matériel through largely hostile territory along treacherous and poorly defended roads, looked for arrangements that would ensure the safety of their soldiers. For this reason, they tried to convince both Moscow and the PDPA leadership in Kabul to make peace with the Tajik commander Ahmad Shah Massoud. The KGB and Eduard Shevardnadze, the Soviet foreign minister, believed that the best way to ensure such an outcome was by showing that Najibullah had the complete support of the Soviet Union, even if its troops were withdrawing. This meant not only meeting all of his requests for matériel, but also being willing to go to battle on his behalf, thus showing his detractors within the PDPA leadership that Najibullah was still top man and demoralizing the opposition. When Najibullah refused and insisted that the Soviet military help to attack Massoud, he had the backing of the KGB as well as of Shevardnadze.

Gorbachev's own position changed several times during this period. In July 1987, he insisted that further Soviet participation in military action was out of the question and that there was no possibility of the troops' withdrawal being delayed. Later in the year, he changed his mind on both counts; by February 1989, he would change it back again. The Politburo seemed to lurch back and forth between contradictory positions. The fate of Afghanistan was far from inconsequential for Gorbachev and other Soviet leaders, and they looked for ways to implement a withdrawal without leaving chaos in their wake. The situation that played out from April 1988 to March 1989 showed that there was still no consensus on what needed to be done.

There were two conflicting forces pulling on Afghan policy between the signing of the Geneva Accords and the end of the Soviet withdrawal in February 1989. The first was the desire to capitalize on the improvement in US-Soviet relations, which seemed on the verge of radical transformation in the spring of 1988. The second was the desire, and the political need, to demonstrate that Moscow could carry out this radical transformation in its relations with the capitalist world without "abandoning its friends" in the Third World. Gorbachev himself maneuvered between various positions and streams of advice. His preferences were dictated first and foremost by his larger foreign-policy priorities and challenges, less so by the developments in Afghanistan.

Making the Best of the Geneva Accords: The Moscow Conference and After

Moscow signed the Geneva Accords, accepting "negative symmetry," in order to end direct Soviet involvement in a long and bloody war. Gorbachev and other Soviet leaders also hoped that withdrawing troops would improve Soviet relations with the United States. The new relationship might then pay dividends in the form of greater US cooperation in enabling reconciliation in Afghanistan. Throughout the withdrawal period (May 1988 to February 1989), Moscow sought to make the most of the Geneva Accords by continuing talks with the United States, trying to press for enforcement of the accords through the United Nations and continuing negotiations with opposition leaders and with Pakistan. At the same time, however, Soviet leaders largely subordinated the Afghan problem to the key goal of building on the Washington summit and improving US-Soviet relations.

As the previous chapter showed, the Geneva Accords were a much weaker agreement than the one Soviet diplomats had been working toward for many years. It did not obligate the United States to stop supplying the opposition via Pakistan, although technically it did bind Pakistan to stop the flow of arms. The accords had a weak enforcement mechanism: a small UN observation force that could take note of violations and pass them on to UN headquarters. Nikolai Kozyrev, the Soviet diplomat who negotiated the accords, wrote that "the legal documents of the Geneva Accords, even if they were not faultless, could, if strictly adhered to, lead to a settlement of the most important foreign-policy aspects of the Afghan problem: the withdrawal of foreign troops from the country, a halt to outside interference in the affairs of Afghanistan, and repatriation of the main body of refugees."[2] Yet even he admitted that in practice, the Geneva Accords, as signed in April 1988, were a face-saving exercise that allowed the USSR to "withdraw its troops in a dignified manner" and continue to support the Kabul regime, as well as soften the negative reaction to the withdrawal from countries such as Cuba and India.[3]

In conversations with Politburo and party colleagues, as well as with foreign leaders, Gorbachev spoke of the Geneva Accords as the first great success of the New Political Thinking. This had several important implications. If he used force now, it could cost him some of the political capital the Soviets had accumulated. If Najibullah fell too quickly, however, it could be ammunition for the conservatives and could harm the Soviet Union's relations with its allies. At a Politburo meeting on April 18, Gorbachev made it clear that fostering the New Political Thinking was more important than worrying about what allies might think: "We have an agreement. There could be changes as the situation develops. But we will not allow ourselves to violate the agreement, especially before the eyes of the whole world."[4] Significantly, he also assigned Aleksandr Iakovlev, the most liberal of the reformers in the Politburo, to the Afghanistan Commission. The Afghanistan Commission, Gorbachev said, had to take advantage of the Geneva Accords and continue "untying the knot of the collision of interests on various levels: the world, the region, and Afghanistan."[5] Assigning Iakovlev to the Afghan Commission reflected Gorbachev's commitment to the New Political Thinking.

In the months after the withdrawal began, it became clear that, in violation of the Geneva Accords, Pakistan was continuing to aid the *mujahadeen*. Soviet and Afghan diplomats filed numerous complaints with the UN office in Kabul. Among the complaints were reports that Pakistan continued to operate training centers, supply weapons to the opposition, and even actively participate in transporting fighters over the border from Pakistan.[6] Gorbachev had three choices: he could halt the withdrawal; he could undertake major operations to knock out *mujahadeen* positions; or he could limit his protests to the diplomatic arena. More than the withdrawal was at stake. As Gorbachev told his colleagues at a Politburo session on April 18, "We have to get the most out of the Geneva Accords. It's not just about Afghanistan. We are taking major steps toward realizing the New Thinking, recognizing a balance of interests, and searching for paths of cooperation."[7] Afghanistan had been one of the major issues impeding improvement in Soviet-American relations since 1979; now the Geneva Accords offered an opportunity not only to remove that obstacle to a new détente but also to provide a model for how the superpowers would settle similar difficult issues in the future.

The behavior of Soviet diplomats in the weeks around the Moscow summit showed the United States that Moscow was looking first and foremost for an improvement in bilateral relations. When the US chargé d'affaires in Kabul, Jon D. Glassman, met the Soviet ambassador, Nikolai Egorychev, the latter avoided any discussion of violations of the Geneva Accords by either Pakistan or the United States. When the US chargé brought up Afghan allegations that the accords were being violated, Egorychev replied: "The Soviet Union works with the Afghan government but is not responsible for its actions. Nor . . . is the United States responsible for the acts of the *mujahadeen*." In his report back to the State Department, Glassman noted that "Egorychev appeared to be dissociating the Soviet Union from RA allegations of Pakistani Geneva violations."[8]

Why was Gorbachev suddenly willing to leave Afghanistan off the table in his relations with the United States? In the spring of 1988, US-Soviet relations were on the verge of an unprecedented breakthrough. The Moscow summit promised to be the culmination of Gorbachev's "peace offensive." Gorbachev's standing and popularity rose in the United States, in Europe, and even at home.[9] The presence of Soviet troops in Afghanistan had been a major obstacle to improving the US-Soviet relationship, and signaling the seriousness of Soviet intentions to withdraw in September 1987 had facilitated the ultimate success of the Washington summit that December.[10] Concerns about sustaining this momentum eclipsed, for the time being, concerns about what might happen in Afghanistan following a Soviet withdrawal. Furthermore, despite the disappointing US attitude on arms supplies in the winter of 1988, Gorbachev still held out hope that eventually the Reagan administration might prove more cooperative, particularly if there were gains in other areas of the relationship. Finally, Moscow would try to rely more heavily on the United Nations in helping to regulate the conflict and limit Pakistani interference.

Moscow's reluctance to let Afghanistan mar the improvement in US-Soviet relations was evident at the Moscow summit itself, in May. Afghanistan was discussed at the experts and foreign-ministers level, along with a host of other regional issues, including unrest in the Horn of Africa and Central America. The Soviet Union and the United States were still far apart when it came to resolving regional problems. Shevardnadze reported at the plenary session that, on each of the topics discussed, "deep and serious issues remain. In a few areas, the method and procedures for a settlement seemed in sight, but further work was required." Yet on Afghanistan, Shevardnadze restricted himself to commenting on Pakistani violations of the accords and the importance of upholding the accords in general.[11]

In the plenary session, Gorbachev tried to push his broader ideas on regional conflicts as well as on Afghanistan. The American side should take him seriously, he said, when he spoke of finding a new way in which regional conflicts could be solved. The US side could be assured that "the hand of Moscow would be a constructive hand." Afghanistan, he told his counterparts, was a "thing of the past," and should be seen as the first example of Third World conflict resolution by the United States and the Soviet Union. But he also urged the United States to help settle the conflict. He did not want to see a fundamentalist Muslim government there, but he would support the transition to a coalition government. Unlike earlier discussions on Afghanistan, here Gorbachev did not accuse the United States of playing an obstructionist role; all of his complaints in this regard were reserved for Pakistan.[12]

While unwilling to press the issue too forcefully on a bilateral level with the Americans, particularly around the time of the May summit, Moscow did bring up violations with UN officials. Soviet and Afghan diplomats sent numerous reports of violations to the UN Good Offices Mission in Afghanistan and Pakistan (UNGOMAP), citing the existence of bases on Pakistani territory, as well as the continued movement of arms across the border.[13] They also made appeals in public and in a confidential manner to the UN officials.[14] In the fall, the Ministry of Foreign Affairs even published a "white book" (detailed report on a specific issue) called *Fulfillment of the Geneva Accords Is in the Interests of All Humanity*. Toward the end of the summer, when the military situation in Afghanistan was becoming particularly difficult, the tone of Soviet protests became harsher. An editorial in the September 1988 issue of the Soviet journal *International Affairs* complained about the "gross violation" of the Geneva Accords by Pakistan: "The Pakistani president pretends that there is nothing worthy of attention in the Geneva Accords except the withdrawal of the Soviet armed forces from Afghanistan. Foreign interference in Afghan affairs did not stop after May 15; it intensified." The editorial even went on to criticize the United States directly: "Nor can we understand the attitude of the United States, one of the signatories of the Declaration on International Guarantees and the Agreement on Interrelationship. Commenting on his talks with Mikhail Gorbachev, President Reagan said that the Soviet Union's decision to withdraw its armed forces from Afghanistan created a positive precedent for settling other regional conflicts. Meanwhile, the United States pursues, through Pakistan, its arms deliveries for the opposition, to the tune of millions of dollars."[15]

The problem for Moscow was that, having committed to withdrawing Soviet troops from Afghanistan, it now had much less leverage in negotiations. As the situation grew more desperate in the late summer and fall of 1988, Soviet officials moved from filing complaints to hinting that the Soviet Union would have to stop the withdrawal. On October 1, at a function for the Chinese National Day celebration, Soviet minister-counselor Botshan-Kharchenko approached Milton Bearden, CIA station chief in Islamabad, and told him that if attacks on Soviet troops did not stop, the withdrawal would be halted. Their exchange, as related by Bearden, illustrates well the attitude of the United States and the difficulty for Moscow in making these sorts of threats:

> *Botshan-Kharchenko:* You must understand, Mr. *Buurdon*, that these attacks against our troops as they withdraw must stop.
> *Bearden:* And if they don't?
> *Botshan-Kharchenko:* Then perhaps we will halt our withdrawal. Then what will you do?
> *Bearden:* It is not what I will do, Counselor; it is what the Afghans will do. And I think they will simply keep on fighting and killing your soldiers until you finally just go home.
> *Botshan-Kharchenko:* But you have some control over such matters.
> *Bearden:* No one has control over such matters, Counselor, except the Soviet Union.
> *Botshan-Kharchenko:* Mr. *Buurdon*, you must still understand that there will be consequences if these attacks continue.
> *Bearden:* I am sure there will be, Counselor.[16]

Since February, US officials had been convinced that the Soviet Union would pull back its troops. This had allowed George Shultz to maintain a tough line on symmetry when negotiating with Shevardnadze in March and to brush off any hints that Moscow might halt the withdrawal.[17] The acting US secretary of state, Michael Armacost, summarized the US position in a conversation with Indian prime minister Rajiv Gandhi in June: "There was no evidence of Soviet suspension of withdrawal. It was hard to see how they could now do so. Forces impelling continuing withdrawal were greater now than they were when the withdrawals had begun."[18]

By September, with the situation in Afghanistan growing more desperate for Najibullah, Moscow tried to regain some leverage vis-à-vis Pakistan and the United States by showing that the withdrawal was not irreversible. Bringing home the troops increased Gorbachev's popularity abroad and at home, but signs that neither the US nor Pakistan was about to give up on supporting the *mujahadeen* exposed him to criticism from those who had been skeptical of the Geneva Accords in the first place. At a Politburo meeting on Afghanistan, Gorbachev agreed that Moscow would have to start taking a harsher line. Rather than saying publicly that the Soviet Union was committed to withdrawal, officials should emphasize that the complete return of Soviet troops was linked to the developing situation in Afghanistan. In other words, if the United States and Pakistan continued to be uncooperative, Moscow might reconsider its commitments under the Geneva Accords.[19] Soviet diplomats cited the accords at every opportune moment.[20] They tried to use the United Nations as a forum to highlight US and Pakistani noncompliance and to underscore the seriousness of the USSR's threats to keep its troops in Afghanistan. Soon after the September 18 Politburo meeting, Shevardnadze asked the UN Security Council to convene a meeting to discuss violations of the Geneva Accords and threatened to delay the troop withdrawal.[21] At the meeting, the Soviet representative charged UNGOMAP with "not doing its job properly," an accusation the United States rejected.[22] Throughout the autumn of 1988, Soviet diplomats would continue to insist that if the situation in Afghanistan were not settled by February 15, Soviet troops would stay beyond the deadline. The bluff failed to work. US diplomats saw such claims primarily as a tactic and had little doubt that Soviet troops would withdraw by the deadline.[23]

Even as the possibility of threatening a continued Soviet presence faded, new opportunities arose for solving the problem through diplomacy. Since the launch of National Reconciliation, Soviet diplomats, advisers, KGB officials, and the military had been engaged in an effort to negotiate with rebel leaders to bring them into a coalition government. With the start of the withdrawal, these efforts intensified. One Soviet Foreign Ministry official even earned the nickname "Mujahed" from his colleagues, because he spent so much time negotiating with rebel commanders.[24] In the summer and fall of 1988, these efforts began to show some success.

Soviet diplomats and others working to open channels to the *mujahadeen* from March 1988 to February 1989 were operating with two goals in mind. They needed to ensure the safety of Soviet troops during the withdrawal. For this reason, they were willing to accept cease-fires that did not necessarily extend to the Afghan army.[25] At the same time, they were also trying to continue the long-term work of forming a stable government in Afghanistan.

By continuing talks with Pakistan, as well as with individual commanders, Moscow was able to take advantage of Islamabad's earlier desire to create a coalition government in Kabul. Pakistan had originally refused to discuss the issue, then demanded that it be resuscitated when the Geneva Accords were about to be signed. Moscow, which had been trying to push a coalition government since the end of 1986, did not want to delay the start of the withdrawal any longer by agreeing to wait for one to be formed. Now that the withdrawal had started, however, Pakistan's interest in a coalition government offered Soviet diplomats a new opportunity.

For Moscow, this development seemed to herald a new opportunity to work for the formation of a coalition government that included the Soviet Union's own allies and moderate opposition elements and that was at the same time strong enough to stabilize the country. Even as Soviet officials conducted talks with rebel leaders, they continued discussions with Pakistan about the possibility of a coalition government. In the summer of 1988, President Zia told Vorontsov, Moscow's ambassador to Kabul, that he would support a solution in which a third of the government would be PDPA, a third would be the "moderate" opposition, including royalists, and a third would be from the "Peshawar Seven." Vorontsov passed the message on to Moscow and received a positive response.[26] Although such an arrangement might face opposition from Najibullah or others in the PDPA, the opportunity to form a government that contained Moscow's allies but was also recognized by Pakistan was too good to pass up.

Zia's death in a plane crash that summer put an end to that particular opening. Other opportunities appeared, however. In December 1988, the UN Secretary General helped to arrange a meeting in Saudi Arabia between Vorontsov and *mujahadeen* leaders, including Barhanuddin Rabbani, head of the Islamic Society of Afghanistan (Jamiat-e Islami), one of the resistance groups based in Peshawar. Although the meeting itself was a sign of how far the Soviet Union was willing to try to find a settlement in Afghanistan, it did not produce any concrete results.[27] The bigger problem for the Soviets in trying to negotiate a coalition government was the continued difficulty of pushing Najibullah and the PDPA toward an agreement. (Another problem was the continued infighting among the "Peshawar Seven" and their ISI interlocutors, which, in the Soviet view, had only gotten worse since Zia's death.)[28] As we will see below, disagreements among various Soviet offices and services, particularly among the KGB and the military, made the effort to press for a coalition government even more difficult.

Moscow's effort to use diplomacy to strengthen the Geneva Accords, from April 1988 to February 1989, brought few concrete results. Soviet diplomats could threaten to suspend the withdrawal, but hardly anyone seems to have taken such threats seriously. In February 1988, it had become very clear to US policy makers that the Soviets wanted out and were unlikely to go through with threats to put off the withdrawal. Gorbachev's desire to build on the improving relationship with the United States made such a possibility even less likely. But his earlier concerns about how the withdrawal was perceived in the Third World and the possible reactions of conservatives at home had not disappeared; he was keen to prove that he could protect Soviet prestige even while engaging with, and making concessions to, its main enemy. When the Geneva Accords were signed, Gorbachev said that they could be a model for the new way of solving conflicts. Thus, Moscow was still looking for ways to protect its interests within Afghanistan and avoid a collapse of the PDPA government. The failure to do this through diplomacy and the United Nations led Gorbachev, in the fall of 1988, to entertain and accept proposals for desperate last-minute offensive measures.

The KGB and Najibullah, the Military and Massoud

For some time, a conflict had been brewing among senior Soviet officials working on Afghanistan. The Soviet military and the KGB had taken sides in the Khalq/Parcham split almost from the beginning of the intervention. The critical situation in the summer and fall of 1988 brought these divisions out. Rival RA leaders tried to take advantage of the differences among Soviet officials to gain advantage. The disagreements of Soviet officials within Afghanistan echoed uncertainty at the Politburo level, whose members were divided about the best course to pursue in Afghanistan. Shevardnadze and Kriuchkov continued to believe that Moscow had to put all its weight behind Najibullah, while others were willing to see a Najibullah-less PDPA enter into a coalition with opposition movements.[29]

One of the biggest areas of disagreement was the extent of support for Najibullah. An area where the split emerged initially concerned the formation of a "presidential guard." The formation of the guard, which was supposed to be loyal to Najibullah alone and provide for the defense of the government in Kabul, reflected how little confidence the Afghan president had in his own military. Soviet military officers did not support the idea. At an August 4 meeting with Shevardnadze during the foreign minister's visit to Kabul, Varennikov argued that the guard was doing more harm than good and upsetting Afghan army officers, who complained that guard officers were earning five to ten times more than they did.[30] With Shevardnadze's support, however, Najibullah was able to continue developing the guard.

Najibullah's support among other PDPA leaders had never been absolute. On the one hand, he was not trusted by Khalqis any more than Karmal had been. His previous tenure as the chairman of the dreaded secret police, or KhAD, likewise did not win him many friends. Key to Najibullah's ability to stay in power had been absolute support from Moscow. With Soviet troops withdrawing, this aura of absolute support had begun to fade. Najibullah's rivals within the PDPA came out of the woodwork and tried to win the support of Moscow.

The most serious challenge to Najibullah's power to emerge in the fall of 1988 was that of Shahnawaz Tanai and Seid Muhammad Gulabzoi. The two had similar backgrounds. Tanai had joined Khalq in the early 1970s, had taken part in the 1978 uprising, and had subsequently risen through the ranks of the military and the party. From 1985 to 1988, he had been chief of the General Staff. In the meantime, he had also been elected a member of the Central Committee, and in 1987 had become a candidate member for the Politburo. That summer, apparently as a result of Soviet insistence, he was made minister of defense.[31] Gulabzoi had taken part in the 1973 uprising against Zahir Shah and had played a key role in the Saur Revolution. He had had a falling-out with Amin, had returned under Karmal, and then had risen to the post of minister of internal affairs.[32]

On September 2, Tanai approached Mikhail Sotskov, the recently appointed Soviet military adviser in Kabul, and tried to get his support to have Najibullah replaced. Tanai told him, "You, Comrade Sotskov, have to understand something else: everything that Najibullah is doing is to save his own regime, that of the Parchamists. And he will continue to do this, whatever it costs him. The main thing is that he hopes to retain power in his hands."[33] Tanai told Sotskov that, at a PDPA Politburo meeting the previous day, he had made arguments in line with the thinking of the Soviets, and with the military command in particular: a coalition government needed to be formed with representatives of all opposition groups; the defense of Kabul and its communications needed to be improved. Now he made a series of points to Sotskov in support of his bid to oust Najibullah:

1. The army is the only real force in Afghanistan. I have 50,000 troops, and Gulabzoi has 30,000.
2. Ahmad Shah will not become president, since he has support only in the northwest. The people know me and support me. Ahmad Shah knows that I have all the power. . . . I have channels to him and can meet him personally, but . . . first, the Soviet Union has to support me and sanction a meeting.
3. Ahmad Shah must know that you are supporting me. The war could be stopped by dividing the spheres of influence.
4. I will be backed not just by Khalqists, but also by honest Parchamists.
5. Real power is needed in Kabul. The [presidential] guard is inadequate.[34]

Tanai asked that this information be relayed to Moscow, but that it not be shared with any other Soviet representatives in Kabul except Varennikov. Sotskov relayed this information in a cipher on the evening of Sept 2.[35]

Tanai did not give up, and tried to open channels to other Soviet military advisers. On September 6 Leonid Levchenko, adviser to the General Staff, brought Tanai and Gulabzoi to see Sotskov. This time, Gulabzoi made the case for their position, saying that Najib could not hold the reins of power and needed to resign. He suggested Mohammed Hassan Sharq as a temporary leader, and spoke of opening direct contacts with Hekmatyar and Rabbani. When Sotskov reported this to Marshal Akhromeev, the latter told him to sit on the information until Gulabzoi and Tanai came to Moscow at the end of September.[36]

Gulabzoi and Tanai's bid for power failed. It received no support in

Moscow, in part because it never had more than the tacit support of Soviet military advisers in Afghanistan and probably had none from the minister of defense, Dmitrii Iazov, while Kriuchkov and Shevardnadze were strongly opposed to challenges to Najibullah's authority. For months, KGB representatives had been aware that a conflict was brewing between Gulabzoi and Najibullah and had tried to convince the former to make peace. Kriuchkov even met with Gulabzoi directly, and Shevardnadze met with both Gulabzoi and Tanai. Shevardnadze urged Gulabzoi to focus on maintaining unity within the leadership, insisting that "if a split occurs, it will be the end of the Khalqists and the Parchamists." What he said next was particularly revealing about Moscow's concerns: "If you suffer a defeat [*porazhenie*], this would be a serious political defeat [*porazhenie*] for the USSR." Gulabzoi insisted that he was a faithful friend of the USSR, but that he did not trust Najibullah.[37] While Tanai was more open to calls for unity, Gulabzoi proved intransigent. Evidently, the rift between him and Najibullah had grown too wide. On October 6, a week after his meetings in Moscow, Gulabzoi was relieved as head of the Kabul garrison. A month later, he was sent to Moscow as ambassador, where he could no longer pose an immediate threat to Najibullah.[38]

Sotskov and Varennikov were sympathetic to Gulabzoi and Tanai, not just because they represented the Khalqi faction and the Afghan military, but because the statements of the two Afghans corresponded to their own reading of the situation. The Soviet military had been arguing that it was necessary to open talks with Ahmad Shah Massoud, even that it would be possible to entice him into a coalition government. With Najibullah rejecting talks and pressing instead for major operations against Massoud, Tanai and Gulabzoi naturally appeared the more attractive partners. The incident had enraged Kriuchkov. The KGB chief resented the military's undertaking of political and diplomatic activity, which he regarded as his own turf. On September 26, Kriuchkov called in Sotskov and berated him: "You are expected to help the army fight successfully, not to engage in politics. Najibullah is supported by our leadership and by Mikhail Sergeevich [Gorbachev]. We and all of the representatives in Kabul need to support Najibullah."[39]

Relations with Massoud became a major point of disagreement between Najibullah and the Soviet command. Varennikov found Najibullah's hatred of Massoud "pathological."[40] Not long before the Geneva

Accords were signed, Varennikov met with Najibullah to discuss measures to be taken in preparation for the withdrawal of Soviet troops. According to Varennikov, Massoud remained the only sticking point. Najibullah insisted that Soviet troops "liquidate" him, because a political compromise was impossible. Varennikov suggested that Najibullah order his own special forces to conduct the attack. The latter replied that they could not handle such an operation.[41] The incident illustrates not only Najibullah's antipathy to Massoud but also Varennikov's considerations at this point. If he attacked Massoud now, prior to the withdrawal, he would be putting his troops at risk in the months ahead, since Massoud might not show restraint after being pummeled by Soviet planes and artillery.

The deteriorating military situation during the first phase of the troop withdrawal and immediately after its completion only made it more urgent for Najibullah as well as his KGB advisers to confirm that Najibullah was in control and to neutralize opposition to him within the RA. This meant being able to show that he could call on Soviet support to attack Massoud. The military saw things differently. In August, Varennikov sent a report to defense minister Iazov, explaining the military's position and offering short- and long-term reasons for avoiding military confrontation with Massoud, and for focusing instead on political efforts. Varennikov argued that since Soviet troops would be withdrawing through areas controlled by Massoud, it was in their interest to remain on the best possible terms with him in the near future. To do otherwise would put Soviet troops in unnecessary jeopardy.

> In our view, accepting the president's proposal to pull the 40th Army into battle with A. Shah [Massoud] could put our troops in a very difficult position during the second phase of their withdrawal from Afghanistan. Undoubtedly, there will be additional major losses, and the organization of their withdrawal in general could be disrupted. At the same time, achieving the main goal—the destruction of A. Shah—is impossible, since it is necessary to know where he is located, and this is impossible because Afghan intelligence has not been able to do this over the past eight years.[42]

Varennikov urged his superiors to focus on reaching an accommodation with Massoud, a strategy that was in the Soviet interest because Massoud had let it be known that he had no particular animus against the USSR and would be willing to maintain contacts. The RA government, he argued, should be willing to accept "any compromise," including granting autonomy to the northern provinces. Varennikov continued to argue that a military operation against Massoud was unadvisable. Aside from being impossible militarily, it would put Soviet troops "in a very difficult position during the second phase of troop withdrawal."[43] An operation against Massoud would not only contradict Varennikov's policy but also complicate preparations for withdrawal and put Soviet troops in great danger.

Varennikov was not the only one who held this view. There was a consensus on this point among the top Soviet brass in Afghanistan, even if the minister of defense in Moscow did not agree. Notes similar to Varennikov's were sent by General Boris Gromov, the commander of the 40th Army who gained fame for commanding its withdrawal, and by General Sotskov. In another memorandum, they complained that Najibullah seemed to have no long-term plan, aside from finding ways to keep Soviet troops involved in the fighting. They asked the leaders in Moscow to make it clear to Najibullah that this was not an option.[44]

Soviet commanders had been using their channel to Massoud to create some stability in the north and to find some arrangement between government-friendly forces and Massoud's fighters. In their view, this would not only further the process of reconciliation, but would also help create a counterweight to the more extremist Hekmatyar. Varennikov drafted a series of proposals, with Tanai's support, for establishing three militia divisions, to be supplied from the USSR, which would collaborate with Massoud but stay loyal to Kabul. Najibullah rejected these plans, and instead pushed for the formation of a special division to take on Ahmad Shah Massoud.[45]

The tenacity of Varennikov and other senior officers in trying to come to some accommodation with Massoud in this period—over the objections of Najibullah, the KGB, and their superiors in Moscow—is testament to the importance they attached to their initiative. Throughout the fall of 1988, Varennikov tried to arrange meetings with Massoud. In one message, he offered seven points for discussion, including the creation of a Tajik autonomous region within Afghanistan, the creation of regular units built around Massoud's forces, and the offer of economic aid not only from the Kabul government but also from the USSR.[46] On three occasions when meetings were set up, however, RA forces carried out attacks in the southern Salang area, forcing Najibullah to call off the talks. Efforts to open a dialogue with Massoud continued, but were undermined by the insistence of Najibullah that the focus should be on military activity against the Tajik's forces.[47]

As in earlier periods of the war, the commanders on the ground had difficulty finding an advocate for their views in Moscow. It was becoming clear that their point of view was losing out to the Kriuchkov-Shevardnadze line. On the morning of September 5, 1988, Sotskov received a phone call from Iazov, who made it clear that the line being taken by Sotskov and Varennikov would not hold:

> Our strategy is to keep Afghanistan friendly. If we keep 50 percent, we have solved our problem; if we run away, they'll come up to the borders of the USSR. Yesterday Mikhail Sergeevich called me in. He demanded that a blow be delivered to Ahmad Shah, against whom you need to be active. M.S. [Gorbachev] is very worried about Ahmad Shah—worried that he's gaining strength but not being hit. The confrontation between us, the Ministry of Defense, the KGB, and the embassy is a conflict he does not support. We have to act together and do everything to ensure that something can remain. You have so many forces there, and yet you're being hit. You have to think about defending Kabul. . . . You have to work with the minister of defense, and you keep talking about Najib.[48]

The Politburo records available at the Gorbachev Foundation Archives, while incomplete, help to give a picture of how these inter-service disagreements played out at the most senior policy-making level. Iazov did not ignore the arguments made by commanders in Afghanistan. At the September 18 Politburo meeting, he spoke for an agreement with Ahmad Shah Massoud, explaining that "he is difficult to beat because he has the support of the population."[49] Kriuchkov did not respond directly to this statement, but complained of the "separatist actions of the GRU" —an open criticism of the military's efforts to reach out to Massoud. Gorbachev also chided Iazov for allowing "individuals in the working group" to pursue policies different from those approved by the Politburo, but agreed that contacts with Ahmad Shah could be attempted. If they did not bring results, however, the 40th Army would have to attack him.[50]

The KGB-military rivalry was nothing new. Najibullah's intrigues merely stoked the flames of this conflict and brought it to the fore. Kriuchkov greatly resented the military's conducting secret talks with the opposition, because he viewed this as an encroachment on KGB territory. According to military officers, he would question officers working under Varennikov, in an effort to prove that the military was seeking a separate peace and was ignoring orders from Moscow.[51] Such inter-service accusations need to be taken with a grain of salt. There was a long-standing rivalry between the KGB and the GRU, and distrust between the KGB and the military as a whole; military officers came to see the Afghan war in particular as the result of foolish KGB adventurism. Nevertheless, there is some truth to the fact that both Shevardnadze and Kriuchkov sought to sideline their opponents on the Afghanistan Commission.

Two casualties of their effort to monopolize control of Afghan policy at this time were Georgii Kornienko and Marshal Akhromeev. Kornienko had been at odds with Shevardnadze on Afghan policy earlier in the year, when Kornienko had gone over Shevardnadze's head and convinced Gorbachev to include a deadline for the withdrawal of troops in his statement on February 8. His removal came in the fall of 1988, when he fought against Shevardnadze's and Kriuchkov's efforts to provide almost unconditional support to Najibullah. According to Kornienko, both he and Akhromeev were ultimately removed from the commission for arguing that Najibullah should cede power. Kornienko and Akhromeev pressed this point in early September 1988 at a "working group of four" meeting, which included Kriuchkov and Vorontsov. Shevardnadze apparently complained to Gorbachev that "Akhromeev and Kornienko were not following the Politburo line."[52] Soon afterward, Kornienko was sidelined from all Afghan affairs and Akhromeev received a strong reprimand from Gorbachev. Both men left their respective posts in November. Kornienko was asked to resign; Akhromeev left of his own volition.[53] The removal of Kornienko and Akhromeev from Afghan affairs silenced the chief voices for a settlement not focused on Najibullah.[54]

With Kornienko and Akhromeev out of the way, this period represented the peak influence of Shevardnadze and Kriuchkov on Afghan affairs. Gorbachev followed their line not only on policy toward Najibullah, but also with regard to the military's efforts with Ahmad Shah. He expressed his dissatisfaction with the military to the Politburo: "We must carry out the line of the Politburo and not adapt it to individuals on the

General Staff or in the working group." At a meeting with Rajiv Gandhi in Delhi on November 19, Gorbachev told the Indian leader: "Our people once tried to undertake something by going around Najibullah. This became the subject of a serious investigation, and we have taken measures to eliminate similar [initiatives.]" This signaled a renewed commitment to Najibullah, one closer to the kind Shevardnadze and Kriuchkov urged. Gorbachev seemed committed to treating Najibullah as a partner: "Najibullah is a figure of high caliber. He is prepared to go far. We will do everything not behind his back, but along with him."[55]

Naturally, in these policy battles Gorbachev's voice was decisive. Why did he side with the more hawkish line of Shevardnadze and Kriuchkov, rather than the one advocated by Kornienko and Akhromeev? One reason is that Gorbachev was acting on his considerations regarding broader Soviet commitments and his own standing within the USSR, rather than on any firm belief that one or the other approach was best for Afghanistan. Kriuchkov and Shevardnadze's arguments appealed to Gorbachev's sense that the Soviet Union had to show its ability to remain faithful to old friends. If the Soviet Union now "abandoned" Najibullah after backing him so enthusiastically for three years, this fickleness could put a strain on Gorbachev's relations with other leaders in the socialist camp.[56] The critical situation in the late summer and fall of 1988 only heightened these concerns.

The second reason has to do with politics at the top of the Soviet hierarchy, in the context of Gorbachev's concessions to the United States. By 1988, Gorbachev and Shevardnadze had their first serious confrontations with the military over arms control. Their decision to pursue a treaty on Intermediate-Range Nuclear Forces independently of an agreement on START and Anti-Ballistic Missiles (a long-standing Soviet position), as well as to include the Soviet SS-23 rockets in that agreement, was made over the protests of the military and Akhromeev. At the time, opponents in the military were kept in line by being warned that they were trying to oppose decisions made by the party leadership.[57] Gorbachev was clearly beginning to feel threatened by the military, and seemed to welcome the opportunity to reassert his primacy in foreign-policy decision making. The confrontation may also have pushed him closer to Shevardnadze and Kriuchkov. Since the latter represented the security forces, his support was key for any initiative that could make Gorbachev vulnerable to charges of ignoring national interests.[58]

Prolonging the Agony

After the first phase of the withdrawal was completed on August 15, Najibullah increased his efforts to have Soviet troops launch a "decisive blow," or else halt their withdrawal and put pressure on the opposition. The Soviet military, reluctant to undertake such operations while it was concentrating on withdrawal, resisted Najibullah's requests and urged him instead to focus on negotiations with leaders like Ahmad Shah Massoud. As the relationship between Najibullah and Soviet military commanders deteriorated, he increasingly sought the help of Shevardnadze and KGB chairman Kriuchkov, which led to a pause in the withdrawal and eventually in the operation against Massoud that Najibullah had been requesting for many months. Gorbachev, initially persuaded by Shevardnadze and Kriuchkov, ultimately abandoned his support for their tactics and reasserted the policy he had followed for the previous two years: focus on withdrawal as a priority, and use diplomatic channels to work for a neutral Afghanistan.

Although willing to support Najibullah by providing massive economic aid and military hardware, Gorbachev for the most part resisted requests for major military operations involving Soviet troops. In the summer of 1988, during the first months of the withdrawal process, Najibullah asked about the possibility of conducting a joint campaign alongside India against Pakistan, with the USSR also providing troops. The military situation had been deteriorating since the start of the Soviet withdrawal in May; rebels had taken over some positions previously held by Soviet troops, and Afghan government troops had been forced back from the Pakistani border.[59] But Gorbachev refused, for now, to consider allowing Soviet troops to resume taking part in offensive operations. He did not want to undermine the political and diplomatic gains he had made when he had signed the Geneva Accords. Thus, Gorbachev rejected Najibullah's request for several major operations in which Soviet troops would participate, albeit in a secondary or tertiary role. The only way such an operation would be possible, Gorbachev told Najibullah, was if "an attack on our [Soviet] troops is committed."[60]

Understanding the split between Soviet commanders and Najibullah is crucial to understanding the halts in withdrawal in 1988 and the military operations undertaken in early 1989.[61] Najibullah did not trust Massoud and resisted negotiating with him. Instead, he sought to bypass his Soviet military interlocutors and, through Shevardnadze, convince Moscow to order major strikes on Massoud's positions. One such strike was ordered in January 1988. According to Liakhovskii, Varennikov and his staff made every effort to show that such an operation was militarily inadvisable, compiling aerial photographs of the snow-covered mountains they had been ordered to attack. A team that included Liakhovskii flew to Moscow on February 14 to present the case against the operation. The operation was set aside, but the episode added to tensions between Najibullah and Soviet military advisers.[62]

Najibullah used every possible channel to force the 40th Army to attack Massoud. In September, he wrote a letter to Gorbachev in which he claimed that Massoud was gaining strength and receiving arms directly from the United States. The text of the letter is unavailable, but the Politburo discussions of the letter give us a general notion of its content. The idea of a major operation was rejected for the time being, although Gorbachev took a more hawkish line. The main focus still had to be on "political settlement and normalization." For the first time, however, Gorbachev signaled that he might consider delaying: "We have to stop saying that we will withdraw no matter what. We must tie the schedule of withdrawal to the current situation." This was to be the new line in talks with Reagan and other politicians, as well as in the press and in the United Nations. Similarly, if Ahmad Shah did not want to talk, then the military should consider operations.[63]

In the fall of 1988, the situation became increasingly desperate for Kabul. As the opposition increased its attacks on Kabul airport and on the Khairaton-Kabul highway, Najibullah looked for new military commitments from Moscow. In October, an entire Scud missile battalion was sent to the outskirts of Kabul to hit resistance positions near the capital.[64] There were calls from Najibullah, via Kriuchkov and Shevardnadze, to halt the withdrawal. The military was opposed, according to Gromov, for three reasons: (1) a halt would have meant that the USSR would be violating international accords; (2) the Soviet military would have had to conduct major operations in the North, endangering the highway; (3) a halt would have made the refugee problem much more difficult.[65] Of the three reasons, it is fair to assume that the second was the most pressing: the security of the highway had been a priority of the military since the start of the withdrawal. Yet again, the voices of Gromov and Varennikov failed to carry the day. On November 5, Moscow suspended the withdrawal.

Some in Gorbachev's inner circle opposed the new line. Cherniaev had believed that his boss simply meant to use talk of halting the withdrawal to put pressure on Pakistan and the United States. He expressed great surprise when, a month later, a request came in for a major operation against Massoud. After all, Cherniaev insisted, Gorbachev had told Shevardnadze: "We will not change our decision regarding the withdrawal; the Afghans have to fight for themselves. . . . Under no circumstances could we *in fact* return to participating in this war." Cherniaev feared that just such a return would be considered. He warned his boss that such operations were not likely to save Najibullah—and that even if they did, the price would be too high.[66] In his memoir, Kriuchkov points out with some anger that Aleksandr Iakovlev, a close friend of Gorbachev's and a member of the Afghanistan Commission, had also argued for timely withdrawal and against new operations.[67] For the moment, however, the voices of Iakovlev and others who argued against the operation faded into the background.

It is unlikely that Gorbachev ever considered anything as drastic as stopping the withdrawal completely. As he pointed out to Rajiv Gandhi, only the "tactical steps," not the "general line" on Afghanistan, had changed.[68] These tactical steps included not only a halt in withdrawal and new supplies of weaponry, but, ultimately, the major operation against Ahmad Shah Massoud that Najibullah had been requesting through Shevardnadze and Kriuchkov for the past year. Gorbachev was still maneuvering between potentially conflicting policies: ending the war and building on improvements in East-West relations on the one hand, and protecting his right flank and trying to salvage something in Kabul on the other. Reversing the withdrawal would almost certainly cause a harsh reaction from the United States and its European allies, as well as from Iran and most of the Muslim world.

Najibullah could still exert influence in Moscow, primarily through his interlocutors from the Kremlin, Shevardnadze and Kriuchkov. The foreign minister and the KGB chief made their last pre-withdrawal trip to Kabul on January 12, 1989. The trip was not unlike Shevardnadze's visit on April 3, 1988, when Shevardnadze had prepared Najibullah for signing the Geneva Accords. At a meeting on December 28, 1988, Gorbachev had already confirmed with his colleagues that withdrawal would resume.[69] Now Shevardnadze had to prepare Najibullah for this decision. Najibullah, however, was still obsessed with Massoud. According to notes from their meeting, Shevardnadze promised Najibullah that he would work for an operation against Massoud. This would be a major strike, not a small operation: "It is clear that no local or limited measures will be sufficient to solve the problem of Ahmad Shah [Massoud]."[70] Shevardnadze carried this request back to Moscow, where he managed to get approval for what would become "Operation Typhoon."[71]

In the winter of 1988–1989, military officers who had long opposed such an operation fought a losing battle to convince Moscow to avoid it. Throughout December, Varennikov and other members of the operating group drafted and sent memos explaining that such an attack now would mostly injure civilians, damage communications with Kabul, make the withdrawal more difficult for Soviet troops, and harm the chances of any future reconciliation.[72] Officers of the 40th Army were unhappy with the order. Some spoke openly of refusing to fight, and even of returning medals they had earned in the war, although none appear to have followed through with this threat. A distraught Varennikov approached Vorontsov, the ambassador, asking what he should do. An order had come down and would have to be followed, but it had upset the officers, who felt it was wrong both strategically and morally. Vorontsov advised Varennikov to carry out the order, but limit the strikes to areas where there were no inhabitants. In the end, the military managed to limit the operation to artillery attacks from the highway itself. This minimized losses to the Red Army and the civilian population, yet resulted in extensive damage to villages in the area.[73] Still, the operation destroyed any chance of a Soviet-sponsored peace between Massoud and Kabul, and further poisoned relations between senior officers on the one hand and Gorbachev, Shevardnadze, and the KGB on the other.

Members of the military had remained unhappy about the operation throughout, and were strongly opposed to the idea of leaving a small force behind. General Sotskov has written of Operation Typhoon: "Almost ten years of the war were reflected, as if in a mirror, in three days and three nights: political cynicism and military cruelty, the absolute defenselessness of some, and the pathological need to kill and destroy on the part of others. Ten years of bloodletting were absorbed into three awful days."[74]

Gorbachev's closest advisers disapproved of his tilt toward the Kriuchkov-Shevardnadze line, and tried to convince their boss that he was making a mistake. Cherniaev sent him a memorandum at the end of October, arguing against a number of developments in Moscow's policy, including the halt of the withdrawal and the planning of an operation against Massoud that was being requested by Kabul.[75] Similarly, Georgii Shakhnazarov warned Gorbachev in December that Najibullah's approaches to Hekmatyar (whom the RA leader preferred to Massoud) were dangerous, because Hekmatyar was an extremist and not likely to compromise. He disagreed with claims that Massoud and Rabbani would be more likely than Hekmatyar to cause problems in Soviet Central Asia.[76] And in January, Vladimir Zagladin, another aide, warned Gorbachev not to order military action against Massoud, since this would hurt any chances of forming a Soviet-friendly coalition government.[77]

In all likelihood, those who saw Najibullah's salvation in a deal with Massoud and those who believed in backing a "Pashtun" government led by Najibullah were exaggerating their case. Massoud was a brilliant commander, an able administrator, and relatively moderate, but he had used previous truces to rebuild his forces and go back to war. Supporters of the "Pashtun" option, on the other hand, were too quick to dismiss the possibility that an influential Tajik figure could take part in a government. Yet what swayed policy at the Politburo level in the fall of 1988 had little to do with these debates about how best to form a government in Kabul. Instead, the crucial factors were Gorbachev's considerations about his standing with other Soviet elites and his concern about how the withdrawal would be perceived. These factors led him to side with Shevardnadze and Kriuchkov, who believed that the Soviet Union's priority was to show support to Najibullah.

Why was Gorbachev, who was generally averse to using force, willing to undertake Operation Typhoon over the protests of many of his advisers and the military? First and foremost, of course, was the desire to leave some sort of stable government in Afghanistan that would not collapse after February 15. Gorbachev had hoped to do so through diplomacy, but by January 1989 it was clear that the Geneva Accords had done little or nothing to advance an internal Afghan settlement. Operation Typhoon may have seemed like a way to give Najibullah additional political breathing space, even though in practice its military value was limited. Gorbachev still believed in the importance of maintaining Soviet prestige in the Third World, which in turn made him more open to arguments by Kriuchkov and Shevardnadze that Moscow first and foremost had to show support to its friend.

Even more important, at this stage, was that Gorbachev—and Shevardnadze—were increasingly worried about protecting their "right flank" internally. Kriuchkov was a crucial supporter, without whom fending off attacks from conservatives would be very difficult. If the Kabul government collapsed soon after the withdrawal—a serious possibility in late 1988 and early 1989—Gorbachev would need Kriuchkov's support against any attacks on his handling of the problem. It thus made good political sense to follow the KGB chairman's advice. Moreover, by this point Gorbachev had allowed Kriuchkov and Shevardnadze to dominate Afghan policy. Their ability to direct information and block opposing arguments was crucial to his understanding of the problem, particularly as demands for his attention grew in the face of mounting economic problems and political difficulties.

Back over the Friendship Bridge

After the "pause" in November and December, Soviet troops continued their withdrawal in January 1989, and were on target to complete it by February 15, the deadline mandated by the Geneva Accords. Yet the situation had not markedly improved for Najibullah in this time. He still faced the same hostile opposition, now emboldened, and a government and military whose loyalty to their president was shaky at best. Moscow now had to decide how to shape its relationship with Kabul: Would further military involvement be possible? How many military advisers could stay behind? These questions were still not settled in the first months of 1989. The crisis of the Najibullah regime in this period forced Moscow to solve them by improvising in response to the situation as it developed.

Just how much Kriuchkov and Shevardnadze believed in the importance of protecting Najibullah's regime became evident when, in the final weeks of the withdrawal, they pushed to have 10,000–15,000 Soviet troops stay behind, guarding the roads and thus providing a life-line for the regime. It was an idea that had originated sometime during February–April 1988. Moscow's decision to sign the Geneva Accords without a US agreement to stop supplying arms to the opposition had put the Soviets in an awkward situation with regard to Najibullah and his allies. Since this had been a key precondition for Soviet and Afghan negotiators in earlier years, it might look like a betrayal, weakening Najibullah in the eyes of both the PDPA and the opposition. It also meant that the armed forces of the RA would be facing an opposition that still received substantial support from Pakistan and the United States, as well as from Saudi Arabia and Iran.

In the remaining weeks before the February 15 deadline, Soviet officials wrestled with the still-unsettled question of how to define the Soviet-Afghan relationship after the withdrawal. As in earlier periods, Gorbachev looked for ways to avoid a chaotic collapse in Afghanistan. In 1988 and the first months of 1989 in particular, he entertained the possibility that some Soviet troops would remain—a position advocated by Kriuchkov and Shevardnadze. On January 24, the Afghanistan Commission submitted a lengthy memorandum that suggested a number of options. It highlighted the numerous difficulties that Najibullah faced, Pakistan's violation of the Geneva Accords, and the importance of maintaining a road link between the Soviet Union and Kabul. The memorandum offered a number of ways that Soviet troops could be kept in Afghanistan to guard those roads, and suggested forming volunteer divisions to carry out the task, offering soldiers a salary of 800–1,000 rubles a month—wages that were unheard-of at the time, even for officers.[78]

The idea to leave some Soviet troops in Afghanistan after the withdrawal had received its latest incarnation around the time of Operation Typhoon, during Shevardnadze and Kriuchkov's last pre-withdrawal visit to Kabul. After meetings with a number of Afghan officers and officials and with Najibullah, Shevardnadze agreed that leaving behind 10,000–15,000 troops was necessary to prevent a collapse in Kabul. At a meeting with officers and staff in the Soviet Embassy, Shevardnadze laid out his proposals.[79] Varennikov and other senior officers were against leaving an exposed division to continue guarding the highway for Najibullah. It was also a question of logistics. They argued with Shevardnadze and his subordinates that it was not possible to leave 10,000–15,000 troops because those soldiers would need support.[80] Once again, members of the military command in Afghanistan were pitted against officials from Moscow.

Gorbachev's more reform-minded advisers were horrified. On January 20, 1989, Iakovlev called Cherniaev, informing him that Shevardnadze was circulating a plan to send 3,000–5,000 Soviet troops to launch a breakthrough attack on the road to Kandahar and act as a convoy for goods. In a conference call a short while after, Cherniaev and Iakovlev pleaded with Gorbachev not to act on the proposal. The conversation also included Shevardnadze, who argued that the USSR had a responsibility to help Najibullah: "You weren't there; you don't know how much we've done over ten years." Gorbachev listened to the discussion, then hung up to call Kabul.[81] The debate erupted again at a Politburo meeting on January 24, when Operation Typhoon was already under way. Shevardnadze insisted that the USSR remained responsible for protecting the Najibullah government: "The fate of the regime is not inconsequential for us. Our Afghan friends ask us not to leave them without support." He went on to highlight the awful state the country was in: "There's already a blockade of Kabul. We are leaving the country in a pitiable state. The cities and villages are ravaged. The economy is paralyzed. Hundreds of thousands of people have died." The withdrawal, Shevardnadze said, "will be seen as a major political and military defeat."[82] No doubt the "emotional" or "personal" factor was at play here—Shevardnadze was genuinely worried that Najibullah and his colleagues, political allies, and perhaps even personal friends, might perish in a bloody confrontation as *mujahadeen* took the city. But he also knew that Gorbachev shared his concerns about allowing the withdrawal to be seen as a defeat, and he dramatized this possibility for the Politburo.

Shevardnadze found himself isolated among his colleagues. No one else seems to have spoken up in favor of keeping troops in Afghanistan. Gorbachev rejected Shevardnadze's arguments, calling his presentation "empty, hawkish babble." As for the fate of Najibullah, Gorbachev said: "We are not going to save the regime. We've already transformed it." Chebrikov, Ryzhkov, and Iakovlev all agreed.[83] The military involvement had to end: "[We] need to hold our principled line, so that we are completely disengaged from their fight," Gorbachev went on. But he agreed with Shevardnadze that the Soviet Union still had an interest in Afghanistan, and that Moscow could not "run away" from the problem. "There are some people, there are comrades, who say: 'So what? We didn't start it!'" Gorbachev rejected these arguments as well: "Capitulation—running away—is foolish, wrong. . . . We cannot appear before the world in just our underwear."[84] As the protocol of the meeting reveals, the Politburo recognized the importance of protecting the Khairaton-Kabul road,

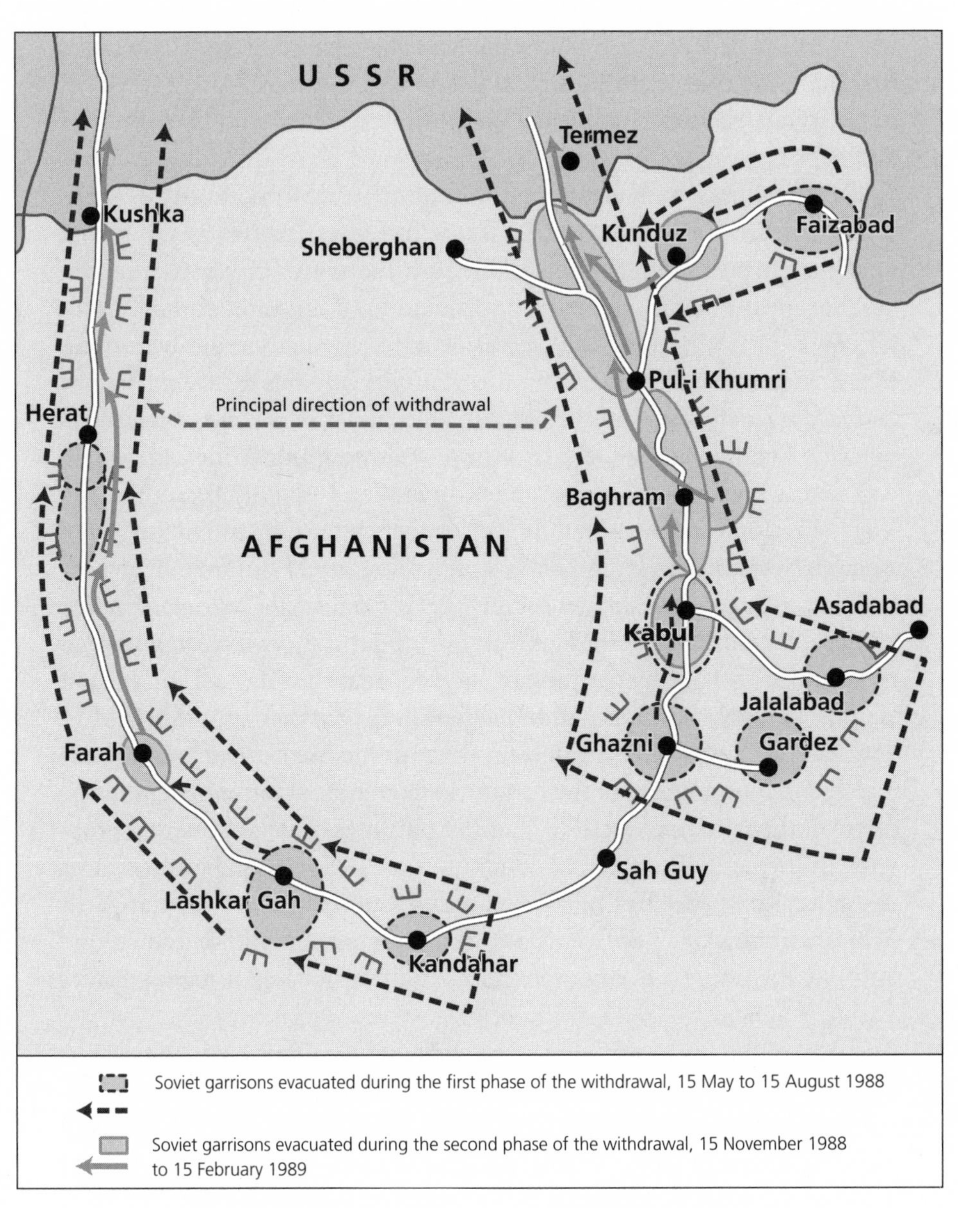

The Soviet withdrawal from Afghanistan

but limited Soviet aid to supplying the Afghan divisions that would be doing the work.[85] Leaving these troops was not just a logistical challenge that the military clearly opposed; it would also mean exposing the USSR to criticisms claiming that it had had no intention of withdrawing all of its troops anyway.

The withdrawal was completed on February 15, 1989, in full compliance with the Geneva timetable, and with representatives of the Soviet and foreign press corps present. Although the pause in November–December meant that more Soviet troops had to return through harsh winter conditions, the withdrawal was effected in good order and with minimal Soviet losses, a testament to the planning and logistical preparation of the Soviet military and the diplomatic efforts of a wide range of officials.[86] The footage of General Gromov dismounting from a tank halfway across the Friendship Bridge to Termez and walking the rest of the way to his homeland was seen the world over. Yet his statement that there were no Soviet troops left behind was not quite true: a number of advisers had stayed in Afghanistan, some of whom would later take part in key battles. Nor was the question of how far support for Najibullah could go in the future settled by the time of the withdrawal. Although the idea of keeping a whole division within Afghanistan had been discarded, Moscow had taken measures to allow for the possible use of Soviet air support and even ground troops in the future. The document formalizing the last phase of the withdrawal stated: "For the purposes of solving sudden problems in the case of worsening condition on the Soviet-Afghan border or in Afghanistan, [we are] making provisions to maintain temporarily, in battle readiness on USSR territory, three motorized rifle divisions, one airborne division, six aviation divisions, and two helicopter regiments."[87]

The start of the Soviet withdrawal from Afghanistan in May 1988 brought the Najibullah regime to a crisis point. The signing of the Geneva Accords emboldened the opposition and frightened members of the People's Democratic Party of Afghanistan. Despite years of training and support, the army of Afghanistan had not proven that it could fight independently. Najibullah's requests for aid in these difficult months were meant to take some of the wind out of the sails of the opposition, and at the same time build him up in the eyes of the party and nonparty figures within the government. As Chapter 3 showed, Soviet confidence in Najibullah helped him to assume and consolidate power in 1986 and 1987, but not to the point where he could truly claim to have a solid independent power base even within his party.

Soviet policy in response to this crisis was dictated by two conflicting priorities. On the one hand, the US-Soviet relationship was improving rapidly in 1987 and 1988. As the previous chapter showed, Moscow saw the Geneva Accords as a result of this improvement and as a catalyst for further détente. This priority was well reflected in Moscow's attitude around the time of the Moscow summit in May 1988, and in the months afterward. Soviet leaders and politicians at all levels kept violations of the Geneva Accords out of discussions with their US counterparts; when the issue was brought up, they were careful to blame Pakistan and not the United States.

The growing crisis, however, forced Moscow to reassess its priorities. Starting in the late summer, key Soviet policy makers began urging the Politburo to approve additional military strikes within Afghanistan. The two most senior supporters of this policy, KGB chairman Kriuchkov and foreign minister Shevardnadze, pressed the idea on Gorbachev. They faced formidable opposition from other policy makers—not just "liberals" like Iakovlev and Cherniaev, but even more cautious reformers like Kornienko. Although Gorbachev resisted similar pleas from Najibullah earlier in the summer, he now increasingly sided with the Kriuchkov-Shevardnadze line. There are several reasons for this.

First, it was becoming clear that Geneva had failed to pay any dividends, as far as the situation within Afghanistan was concerned. When he had signed the accords, Gorbachev still hoped that the withdrawal itself would prompt the United States to stop supporting the *mujahadeen* and would make a settlement which included the PDPA more favorable. In practice, the blatant violation of the accords by Pakistan was an embarrassment, since it revealed that the accords were really little more than a fig leaf.

Second, concerns about the effect of the withdrawal on other Soviet Third World allies had not gone away. Indeed, at a time of profound changes in Moscow's international relations, Gorbachev had to maneuver very carefully to avoid upsetting the entire system of relations in the Soviet sphere. When Soviet officials visited Vietnam in August 1988, Vietnamese leaders expressed concern over not only the withdrawal from

Afghanistan, but also the Soviets' new receptivity to China. Similarly, the cautious Soviet opening to South Korea worried North Korea.[88] We know that other countries, such as India and Cuba, had made their concerns about a hasty Soviet retreat known earlier. While the limited archival access makes it difficult to assess precisely how widespread this sentiment was among Soviet allies, it is clear that this concern increased during the withdrawal period.

The withdrawal also coincided with a pivotal phase of perestroika and a difficult period for Gorbachev politically. On the one hand, he had been successful in packing the Politburo with his own allies and removing most of the hard-liners he had seen as a threat. His foreign policy, particularly vis-à-vis the West, seemed to be showing significant results. On the other hand, a worsening economic situation was causing discontent in the Soviet population even as nationalist movements were increasingly making their presence felt. The extent of growing opposition to perestroika within the party was brought home to Gorbachev in April 1988, with the publication of a letter by a Leningrad schoolteacher entitled "I Cannot Part with My Principles," in the newspaper *Sovetskaia Rossia*. The Gorbachev described by historian Vladislav Zubok was also highly aware of the fragility of his position and his reforms.[89] His shifts from dovish to hawkish positions and back again, described in this chapter, presaged and echoed similar zig-zags in other areas of foreign and domestic policy, particularly in 1990–1991.[90]

Third, the events described in this chapter highlight the problems of Soviet foreign-policy making in general, and concerning Afghanistan in particular. Splits between Soviet officials in Kabul, particularly the military and the KGB, translated into policy battles fought in Moscow. All sides involved felt that they were acting in the best interests of the Soviet Union, but their policies were often incompatible. The military's relationship with Massoud was incompatible with the KGB's goal of supporting Najibullah, since the KGB believed that Najibullah should be allowed to choose the people he made alliances with. What is striking, moreover, is that these debates were not aired fully in Politburo meetings but were decided, as in many instances during the war, *kuluarno*—that is, in some sort of informal framework. Thus, phone calls or private conferences among Shevardnadze, Kriuchkov, and Gorbachev often predetermined the results of Politburo decisions. At the same time, the sidelining of officials like Kornienko proved an effective way for Kriuchkov and Shevardnadze to increase their dominance in the process of making foreign policy. Thus, after a brief period in which Moscow tried a more "democratic" approach to policy making on Afghanistan—a period marked by long, heated debates involving the full Politburo—a familiar pattern emerged. It resembled the approach taken in the early 1980s, when Andropov, Ustinov, and Gromyko essentially set Afghan policy among themselves.

7

SOVIET POLICY ADRIFT

The withdrawal of Soviet troops from Afghanistan, completed on February 12, 1989, left many questions about Moscow's future policy in that country unresolved. In the years after Soviet troops departed, Najibullah would make requests for Soviet military intervention during several crises. Though all of these requests were ultimately denied, some officials in Moscow believed that such interventions should not be ruled out. Long after the Soviet government had lost the ability to support Najibullah materially, it continued to insist that he be included in any transition government.

In many ways, Moscow's policy toward Afghanistan followed the same basic principles after the withdrawal as it had before. The main goal of Soviet policy was still to prevent a collapse of the Kabul government and to show that it had achieved a controlled transition there. Despite the growing financial and economic crisis within the Soviet Union, military and material aid continued to arrive in Kabul on Soviet transport planes. Soviet advisers stayed behind with the Afghan army. And Soviet diplomats continued their efforts to negotiate for recognition of the regime and for the cutoff of US arms supplies to the opposition.

Following the Soviet withdrawal, internal developments in Afghanistan continued to be important to Soviet leaders. First, Gorbachev's reforms and outreach to the West were no longer going unchallenged. Although nobody in the leadership opposed the withdrawal of troops from Afghanistan, many, like KGB chief Vladimir Kriuchkov, were skeptical about the new multilateral approach to foreign relations and thought that Gorbachev was going too far. Bringing the war in Afghanistan to a peaceful conclusion that allowed Moscow's client to play a key role would help to neutralize their criticisms, particularly if it was done with the cooperation of the United Nations, the United States, and Pakistan.

Second, Afghanistan continued to be important after 1989 for some of the same reasons it had been important in 1979. It shared a 2,000-

kilometer border with the Soviet Union. Even if fears that Afghanistan might become a US base had diminished somewhat, rising nationalism and anti-Soviet sentiment in the republics bordering Afghanistan gave Soviet leaders new reasons to work toward stability there. And last but not least, the war was now too far toward the forefront of the public consciousness to be ignored. Not just Gorbachev but even reformers like Iakovlev worried about the political fallout if the war were declared meaningless—which was likely to happen if Moscow declined to play a further role in events there.

Nevertheless, soon after the withdrawal Soviet policy on Afghanistan was adrift. This is not surprising, considering the escalating crises that Moscow faced from 1989 onward in the domestic and foreign spheres. The collapse of Eastern European communism in 1989, the secessionist movements that grew in strength from 1990, and the challenge of trying to transform the Soviet state now loomed as larger, more threatening, and more immediate problems than the war in Afghanistan. Although preventing a collapse in Kabul was still important to Moscow, the need to guard the Soviet Union's reputation as a defender of global national-liberation movements decreased as the country's superpower status evaporated at the end of 1989.

Gorbachev's greatest success with regard to the withdrawal was on the propaganda front. The withdrawal was welcomed by the Soviet public, which largely credited Gorbachev with ending the war. Moreover, a carefully controlled propaganda campaign allowed him to criticize the war without facing questions about why it had dragged on long after he came to power. Moscow continued to support Najibullah, even as it sought a diplomatic solution to the conflict. This was partly because until August 1991 there were still people in the Soviet government who supported propping up the Najibullah government with supplies. Even after the failed August 1991 coup by conservatives in the party and security services (among them Kriuchkov and Varennikov), the dying Soviet state continued to back Najibullah politically.[1]

Setting a Precedent for Nonintervention: The Battle for Jalalabad and After

If Moscow had a clear Afghan policy in February 1989, we have no evidence of it. Indeed, it seems that Soviet leaders were unsure to what extent they needed to continue supporting Najibullah. Shevardnadze and

Kriuchkov continued to argue for active support for the Kabul regime, even to the point of using Soviet air support during critical battles. Gorbachev himself seemed less interested in propping up Najibullah, and was too preoccupied with other problems to give Afghan policy much direction. As a result, Afghan policy continued on sheer inertia, with the KGB playing the most active role.

Soviet policy makers may have expected events to resolve the issue. The CIA had been predicting that the Najibullah regime would not outlast the Soviet presence in Afghanistan, and their lack of confidence was shared by their Pakistani counterparts. Similarly, some in Moscow, such as minister of defense Dmitrii Iazov, did not think the regime would last more than a few months.[2] As in 1980, the spring thaw, the traditional start of the fighting season, would reveal the relative military strengths of the opposition and the government.

The first test of the Soviet resolve to refrain from further direct military involvement came within two months of the withdrawal. US and Pakistani officials were confident that Najibullah would fall quickly once Soviet troops withdrew. On March 6, 1989, Benazir Bhutto, the recently appointed prime minister of Pakistan, met with the "Afghan Cell," a group of senior ISI and military officers, to discuss the next steps. Her advisers, in particular ISI chief Hamid Gul, urged a frontal attack on Jalalabad. Although the city was heavily defended by government forces, Gul was confident it would fall within twenty-four hours. US ambassador Robert Oakley, who was present at the meeting, believed Gul was right. The plan was approved, and soon the CIA began directing Toyota pickup trucks and weapons to positions outside Jalalabad.[3]

Several days later, the attack began. It was the first major attack since the withdrawal, and hundreds of young Afghan men and boys recruited from refugee camps took part. In Kabul, Najibullah grew nervous. Several urgent telegrams were sent to Moscow. Jalalabad, Najibullah said, was about to fall, and then the road would be open to Kabul. He requested support in the form of air cover and bombardment delivered by pilots flying in from Soviet bases.[4]

Gorbachev called together a meeting of Politburo members and Central Committee secretaries at the Novo-Ogarevo dacha to discuss possible responses. Once again, his colleagues took familiar positions. The main supporters of a Soviet intervention were Shevardnadze and Kriuchkov. And it was Shevardnadze, as before, who framed the question in terms of loyalty, saying that if the Soviets let Najibullah fall, this would be a betrayal, and that the Soviet Union's friends in the Third World would see it as such. Iazov, the defense minister, was more cautious, saying that the air bombardment would accomplish little and that Soviet involvement would be impossible to hide. Apparently, it was Iakovlev who spoke most forcefully against intervention, though he had the support of others at the meeting. In the end, Gorbachev said that he was "categorically" against any Soviet bombardment of Jalalabad.[5] Although the Politburo approved additional supplies and the organization of a special supply train, Soviet bombers did not come to Najibullah's rescue.[6] The behavior of Shevardnadze and Kriuchkov demonstrated that the "Najib lobby" would keep working to ensure Moscow's support long after Soviet troops had gone home. If they could not involve Soviet troops, they would at least make sure he had all the means to keep fighting on his own.

According to Cherniaev's diary entry on the meeting, Gorbachev had declared himself categorically against using Soviet pilots to defend Jalalabad, because he didn't want to go back on a promise the Soviet Union had made before the world. This is disingenuous. Gorbachev had proven on numerous occasions in the previous nine months that he believed the Soviet Union had to continue supporting the Najibullah regime or risk undermining its own authority. He had authorized Operation Typhoon, even though there was strong opposition to it on the grounds that it made the USSR look untrustworthy; and he had entertained the idea of leaving troops behind after the withdrawal, in contravention of the Geneva Accords. Yet on this occasion he sided against Kriuchkov and Shevardnadze, thus keeping Soviet troops out of direct involvement.

Gorbachev's decision not to send in Soviet pilots was largely consistent with his attitude prior to the later summer and early fall of 1988; the Soviet Union would continue to support the Republic of Afghanistan with material aid and diplomacy, but not with any military involvement. Gorbachev must have realized that his reaction to Jalalabad would set a precedent—that if Moscow sent in pilots, it would be expected to do so on every occasion when the Najibullah government felt it was in immediate danger. The war would then drag on indefinitely, and Gorbachev would no longer be able to take credit for withdrawing troops. Furthermore, his decision offered a compromise between those arguing for direct Soviet involvement and those arguing for no involvement whatsoever. Whereas Moscow had ceased supplying the Afghan military from the time of the withdrawal, it now committed itself anew to providing military matériel. This, too, set a precedent: Najibullah's military continued to receive Soviet arms until the end of 1991.

Supporting Najibullah and the PDPA and including them in a future government was also beginning to lose some of its importance for Gorbachev. Emboldened perhaps by the positive public response to the withdrawal, he was less worried about the potential political damage he could sustain from not supporting Najibullah. At a Politburo meeting on March 23, 1989, he said that the actual composition of the government in Kabul no longer carried great importance for the Soviet Union: "For us, the main thing is that a hostile government should not take over [in Kabul]. As for the rest, . . . let it be any governing combination—that's not our problem."[7] In practice, however, he never quite made this sentiment official policy.

Najibullah survived the Jalalabad attack. The *mujahadeen* recruits proved unequal to the task that the CIA and ISI had confidently predicted would be accomplished. Massoud kept his men out of it: his officers had assessed the plan and the government's defenses, and had decided that the whole plan was foolish. Suspicion of Gulbudin Hekmatyar and his motives may also have played a role. The RA troops in Jalalabad, after being forced to give up several border posts in the first hours of the battle, held their ground over many weeks. Not only did the Afghan military perform better than expected, but Soviet military matériel, some of it turned over during the withdrawal, overpowered the attackers. The Afghan air force, flying Soviet planes, successfully carried out the bombings Najibullah had asked Moscow to conduct. And Soviet advisers helped to operate the Scud missiles that had been turned over to the Najibullah regime during the withdrawal.[8]

The Jalalabad attack changed the shape of the conflict. In March, Najibullah had appealed to Moscow in panic, but it was a confident and emboldened leader who spoke at the eleventh anniversary of the Saur Revolution. The defense of Jalalabad, he told party members, "can be considered a strong blow to those who were predicting the collapse of our revolution."[9] The foreign press reported that the *mujahadeen* were showing themselves a divided fighting force, incapable of taking full advantage of the heavy arms provided by the United States. There were reports of growing resentment over Islamabad's involvement and its support for the most radical groups.[10] Najibullah's ability to continue governing without the presence of Soviet troops began to impress many Soviet advisers still involved in Afghanistan. The Jalalabad victory emboldened not only Najibullah, but also many of his supporters in the KGB and among Foreign Ministry advisers in Kabul. General Makhmut Gareev, a military theoretician sent to serve as the senior Soviet military adviser in Afghanistan, wrote that by the end of 1989, there was increasing talk of moving the Afghan army from defensive to offensive operations.[11]

Gareev did not share the optimism of these advisers. Najibullah and the regime were still very vulnerable. The budget deficit grew as the economic links of the government frayed and it came to rely increasingly on cash payments, including those made to militias, for maintaining loyalty.[12] More desperate than ever for Soviet support, Najibullah asked for more arms and for Moscow to take a stronger line on Pakistan—to at least consider demonstrative flights over that country's territory. He threatened to take the initiative himself, writing to Moscow: "On our side, we are considering the question of rocket attacks against targets on Pakistani soil."[13] Although we have no record of Moscow's response, the request must have made most Soviet officials bristle. They could not but be horrified by the thought that the war might expand at a time when the USSR could barely afford its already shrinking commitments to Kabul, and when Soviet leaders were looking to their former Cold War enemies for economic aid.

The withdrawal of Soviet troops removed Afghanistan from Gorbachev's day-to-day agenda. In this context, the policies and views of individual officials gained increasing importance. There was disagreement not only over the USSR's future role in Afghanistan, but also over the prospects of the Najibullah regime. These were evident when General Gareev, appointed to take over as the senior Soviet military representative in Kabul, went for pre-departure briefings with senior officials in Moscow. According to Gareev, Dmitrii Iazov told him: "Go there [to Kabul] for two to three months, and then we'll see"—meaning that he did not expect the Najib government to last much longer.[14] Kriuchkov, on the other hand, urged him to work hard to continue supporting Najibullah.[15] Nor had the KGB-military rivalries, discussed in the previous chapter, been resolved. In his meeting with Gareev, Kriuchkov harped on the importance of developing a good working relationship with Najibullah, as well as with the KGB representative in Kabul. When Gareev suggested that Moscow needed to push Najibullah toward improving his relationship with his own army, Kriuchkov showed his displeasure and criticized the "improper position" taken by "certain military advisers working in Kabul."[16] The infighting between the military and the KGB had not ended with the withdrawal.

Whatever strengths Najibullah had shown after the Soviet withdrawal, he still faced bitter divisions within his own government. For one thing, the troubled relationship with defense minister Shahnawaz Tanai had not improved. By February 1990, Tanai had emerged as the leader of a plot to oust Najibullah. As in 1988, Tanai tried to win Soviet support for his plans. This support was not forthcoming. Like General Sotskov, Gareev found himself being approached as a messenger to Moscow, as well as a mediator in the intra-PDPA rivalry. Tanai tried to convince him that the Soviet Union was making a mistake in putting so much support behind Najibullah and the Parcham wing. Gareev could do little besides try to mediate some sort of truce between the two sworn enemies. By January 1990, the conflict within the government had grown so acute that, Gareev found, most of the leadership had given up governing and were devoting themselves full time to infighting.[17]

Najibullah's behavior did not help matters. He relied on his supporters in the state security apparatus to try to weed out plotters, ordering the arrest of 137 army officers he believed might be loyal to Tanai. When Gareev pointed out that the arrests might provoke Tanai, Najibullah rejected the suggestion. Gareev told the president that he was relying too much on the Ministry of State Security, urged him to be more inclusive with Khalqists, and pointed out that in some ways Tanai had a legitimate gripe. Gareev even suggested offering Tanai another promotion and then sending him to Moscow for training, but Najibullah did not believe Tanai would go. By early March, the atmosphere had become so tense that meetings of the commander-in-chief's staff had been abandoned; Tanai was refusing to enter the president's residence, for fear of arrest. Nor did he agree to Gareev's request to meet with Najibullah, saying that such meetings were useless and that he would make no more compromises.[18]

Moscow largely remained aloof from this conflict. When Tanai told Gareev that he and his supporters "were ready to deliver a blow," Gareev passed on the information to Iazov. The latter said he believed he could get Tanai and Najib to meet if he were in Kabul, but he never made the trip.[19] Although reports on the situation within the PDPA would probably have been available at Politburo meetings in Moscow, we have no records to indicate what discussion they sparked or which options were considered. The position of Kriuchkov and the KGB advisers during this episode is, unfortunately, likewise unknown.[20]

In the event, Tanai launched his bid for power on March 7, 1990, ordering jets to bomb Najibullah's palace. Meanwhile, forces loyal to Tanai tried to open a path for Hekmatyar's fighters. As it emerged later, Tanai's plot involved more than the removal of Najibullah. As in 1988, he believed that he could reach out to the opposition and form a new government. This time, however, rather than appealing to Massoud, he had made contacts with Hekmatyar. As the journalist Steve Coll has shown, this was part of a planned "double coup," funded in part with Osama bin Laden's help, to remove both Benazir Bhutto in Islamabad and Najibullah in Kabul.[21] The bombs did not kill Najibullah, and the coup failed within several hours. Units loyal to Kabul routed the defecting units, and Hekmatyar's force never entered the capital.[22] Massoud not only sat this one out, just as he'd sat out the attack on Jalalabad in March 1989, but apparently told his contacts in the Ministry of Defense do to everything possible to quash Tanai's rebellion. Massoud's hatred of Hekmatyar had once again helped to save Najibullah.[23]

Available evidence suggests that the possibility of using Soviet planes to help quash the rebellion was at least briefly considered in Moscow. Several Politburo members spoke in favor of such an operation, citing familiar arguments. Lev Zaikov, a CC CPSU secretary responsible for the military-industrial complex and a Politburo member, said: "If Najib falls, people will say, 'What did we fight for?'" Yet even Shevardnadze was now opposed to using Soviet planes. Najibullah's request was denied.[24]

The Tanai rebellion marked the last time that Soviet leaders seriously discussed the possibility of using the Soviet military to support the Najibullah regime. Indeed, by the fall of 1990 Najibullah's supporters in the KGB were becoming concerned that aid would eventually be cut off entirely by the Council of Ministers, where both the ability and the willingness to continue any sort of foreign aid was eroding. Leonid Shebarshin tried to make Najibullah aware of this trend through the KGB's representative in Kabul. Meanwhile, he and Kriuchkov agreed that they would try to push through as much aid as possible before the tide turned completely against supporting Najibullah.[25]

Although, like the CIA and ISI officials who supported them, the resistance leaders may have felt that the Najibullah regime would be easily toppled, the prospect of victory did little to unite them. On the contrary, forces loyal to Hekmatyar and Massoud were engaged in sporadic fighting by mid 1990, foreshadowing the civil war that would erupt between their forces after the Kabul government fell.[26] The rivalry between the two commanders was made worse by the preference of the Pakistani ISI for Hekmatyar, to whom it directed a much larger share of funds. Hekmatyar showed that he was not above making separate deals with major defectors from the communist government, as he did when he plotted with Shahnawaz Tanai to oust Najibullah in March 1990.[27]

Officials in the KGB and the military now began to worry about the effects of events in Afghanistan on the USSR's Central Asian republics. Although not a great concern in the early years of the war, in 1989 and 1990 separatist and Islamist groups were beginning to make their presence felt, particularly in Tajikistan. Gareev, Kriuchkov, Shebarshin, and others were worried about the potential of a spillover effect if extremists did come to power in Kabul.[28] The possible spillover became an argument for those who wanted the USSR to continue supporting Najibullah politically and economically. In a Politburo memorandum following a trip to Afghanistan in August 1989, Kriuchkov and Shevardnadze pointed out that their conversations with party leaders in Uzbekistan confirmed that "Islamist fundamentalist" groups there and in other Central Asian republics were waiting to take advantage of a *mujahadeen* victory in Afghanistan.[29] Nevertheless, it does not seem that any particular response was discussed at the Politburo level.

The Tanai coup attempt highlighted the fact that Afghan policy was adrift in Moscow. A few weeks later, Cherniaev wrote a memo to his boss suggesting that he request some policy reviews, involving various experts, that would point to a "fundamental redevelopment" of Soviet policy toward Afghanistan.[30] Such reviews did take place through the summer of 1991. Yet none of them led to a fundamentally new policy. In the context of the state collapse within the USSR itself in 1990 and 1991, this is hardly surprising. Furthermore, in March 1990 the Politburo commission on Afghanistan was dissolved. This eliminated the one senior body with coordinating capacity, and at the same time served to increase Kriuchkov's dominance of Afghan policy.[31]

Najibullah's ability to hold on to power after Soviet troops left was a testament to his own political skills, to the work of Soviet advisers who had trained the Afghan military and militias, and to the weakness of the *mujahadeen.* His survival proved to be a political boon to Gorbachev, who did not have to face the political fallout of a bloody collapse in Afghanistan. It also posed a challenge: as long as he stayed in power, Moscow had some obligation to keep its promises and offer him support in the form of aid, advice, and representation on the world stage. Key officials who had been instrumental in steering Afghan policy in earlier periods, including Shevardnadze and Kriuchkov, continued to exert enormous influence and press for a more active involvement. Although they failed to secure ongoing military involvement, they succeeded in ensuring that supplies would continue to flow to Kabul. While Gorbachev accepted their insistence that military and economic aid should continue, he rejected the possibility of allowing any Soviet troops to return. Sending the Soviet air force into Afghanistan could have compromised the political gains of the withdrawal and Gorbachev's new reputation as a global peacemaker, increasingly the only source of his waning popularity.

Continuing Diplomatic Efforts: Pakistan, the United Nations, and the United States

Although the Geneva talks had culminated with the signing of the accords in April 1988, Moscow continued to look for ways to steer the situation in Afghanistan toward some sort of acceptable resolution. While the increasing dialogue with the United States, Pakistan, and Iran seemed to provide new opportunities for a diplomatic solution, Soviet leaders and diplomats were now forced to operate in an arena where they had much less leverage and where Soviet power in general was in rapid decline. This decline made multilateralism all the more crucial, for it offered the only real possibility of protecting Soviet interests in Afghanistan in a way that was consistent with the New Political Thinking. It was also part of Gorbachev's drive to show his conservative critics that he had not abandoned Soviet interests by moving away from confrontation with the USSR's recent enemies.

US policy on Afghanistan was itself in a state of flux after the Soviet withdrawal, which coincided roughly with the coming to office of George H. W. Bush, former head of the CIA and Reagan's vice president. For some officials, particularly the more technocratic ones, the outcome of events was not crucial now that Soviet troops had departed. The US undersecretary of state, Michael Armacost, later reminisced: "As far as I was concerned, we weren't that interested in what happened in Afghanistan internally. We were just interested in getting the Russians out."[32] Still, there were those who continued to believe that nothing short of the complete defeat of communism could be the goal in Afghanistan, and that the *mujahedeen* were the best proxy to win that fight. When the British journalist Anatol Lieven, who had covered the war from Pakistan and from within Afghanistan throughout the 1980s, criticized the US policy of supporting Hekmatyar to a US diplomat newly arrived in Islamabad in May 1989, the latter insisted that the resistance would build a "successful free-market democracy," that destruction of the regime was necessary for the defeat of communism, and that "the Russians did it to us in Vietnam, and we're going to do that to them in Afghanistan."[33]

On the whole, though, the Afghan problem diminished in importance for the United States after the signing of the Geneva Accords and the withdrawal of Soviet troops. The administration of George H. W. Bush was never as ideologically committed to the Afghan resistance as Ronald Reagan's had been. Nevertheless, Bush and his advisers felt they could not support a settlement that allowed Najibullah to stay in power, nor could they cut off aid to the resistance while the Soviet Union continued to support its client in Kabul. Even as US-Pakistani relations cooled, the Afghan effort continued to receive CIA support. Despite growing doubts among some CIA and State Department officials that the US policy of allowing the ISI to channel support to the more extremist elements in the resistance would hurt long-term US interests, the policy did not change substantially.

In Pakistan, the tumultuous domestic situation following the death of Zia ul-Haq helped to make Afghan policy all the more amorphous. Benazir Bhutto, prime minister from 1988 to 1990, generally followed the lead of the ISI on Afghan questions, which in turn led to the disastrous attack on Jalalabad in March 1989. But the ISI itself would later try to use the *mujahadeen* and an anti-Najibullah coup attempt in Kabul to oust Bhutto.

From the signing of the Geneva Accords onward, Gorbachev repeatedly expressed his belief that Moscow could resolve the Afghan conflict through multilateral diplomacy and the United Nations. Pakistani leaders also expressed interest in a greater UN role. Having failed to get a coalition government set up before the Geneva Accords were signed, Pakistan had to be content with a mechanism that allowed the United Nations to stay involved. Yet the UN Secretary General was reluctant to involve his organization in the conflict any further.

At a meeting with Javier Pérez de Cuéllar on the day of the signing ceremony, the recently appointed foreign minister of Pakistan tried to push the UN Secretary General to commit his organization to a continued role in Afghanistan. The accords themselves, he said, "would not lead automatically to peace in Afghanistan. That could only be achieved with the formation of a transitional government. All the Geneva parties were in agreement that Mr. Cordovez should continue his efforts toward reaching an understanding to that end with all the parties concerned." Pérez de Cuéllar was unwilling to commit the United Nations again because the Geneva process itself, which had taken the better part of the decade, had brought his office under fire from a number of quarters. He insisted that it would be difficult for the United Nations to get involved, since the organization was "enjoined from interfering in the internal affairs of member countries."[34]

Moscow offered its own proposals before and after the withdrawal. In New York for the UN General Assembly in December 1988, Gorbachev met with Pérez de Cuéllar. Moscow was ready to accept a neutral government, Gorbachev told the Secretary General, but to ask Najibullah to step down ahead of time was unfair. Later, speaking before the assembly, Gorbachev proposed a cease-fire that would begin on January 1, 1989, a halt of arms supplies to both the government and the opposition, and the "deployment of UN peacekeeping forces in Kabul and other 'strategic centers.'" His proposals were rejected by virtually every country.[35] Another Soviet proposal in March that called for a group of experts representing the United States, the USSR, Pakistan, Iran, the Najibullah government, and the *mujahadeen* also went nowhere.[36]

If during most of the war Soviet leaders had often sought to keep the United Nations at arms length, from 1989 to 1991 they increasingly sought UN help with reaching a resolution in Afghanistan. As leaders in Moscow felt their own ability to influence events in Afghanistan slipping, their calls for UN involvement became more strident. In February 1989, Gorbachev tried to appeal directly to US president George H. W. Bush to accept the idea of an international conference proposed at the United Nations two months earlier. Bush refused.[37] During the Jalalabad battle in March, a Soviet representative delivered an angry message to Pérez de

Cuéllar, accusing him of not taking an active role in Afghan affairs and of allowing the situation to slip out of control.[38] Soviet officials continued to push for an international conference under the auspices of the UN Secretary General, for an increase in the size of UNGOMAP, and for the Secretary General and his representatives to play an activist role for the next several years.[39]

The turn to the United Nations reflected Gorbachev's own faith in that body, and his growing belief in the importance of broad international agreements. It also reflected Moscow's growing realization of its own impotence to dictate the course of events. Moscow counted on the United Nations to help build international support for the Soviet position in negotiations; this in turn might force the United States and Pakistan to modify their demands, particularly on Najibullah. Moscow also counted on the United Nations to enforce those parts of the Geneva Accords that protected the Kabul government—something that (as we saw in the previous chapter) the United Nations was unable and unwilling to do. Finally, Gorbachev clearly hoped that the United Nations would help to provide legitimacy for the withdrawal and for his handling of Afghan policy. As he reiterated on numerous occasions, success in Afghanistan would prove that the New Thinking could combine improvement in East-West relations with the protection of national interests.

Of course the United Nations was not the only venue in which Moscow could seek a solution to the Afghan problem. Afghanistan continued to be on the agenda of bilateral US-Soviet meetings. But efforts over the following months to convince Washington to stop or at least reduce its support to the *mujahadeen* were generally fruitless. In May, Shevardnadze made yet another in a series of private appeals to US leaders during a private dinner. He even went beyond earlier Soviet positions, uncoupling the cessation of Soviet arms supplies to the Sandinistas in Nicaragua and US aid to the *mujahadeen*. For the first time, he even suggested that the Soviet Union would not insist on keeping Najibullah in a coalition government after the settlement.[40]

Gradually, however, the attitude in Washington became more promising. On the one hand, Pakistani officials and their allies in the US Congress still opposed the idea of a transitional government that included Najibullah.[41] On the other hand, there was a growing consensus that the United States had reached its main objective (the Soviet withdrawal), and the realization was emerging that continued support for the *mujaha-*

deen might not be in US interests. Whereas in late 1988 and early 1989 CIA analysts were confident of a quick military victory, the RA army's successful defense of Jalalabad in the spring of 1989 seemed to change the calculus.[42] Furthermore, now that Soviet troops had left, US officials and leaders began doubting whether a military takeover of Kabul would even be in US interests. In August, UN officials had learned that both the United States and Pakistan were "reevaluating the desirability of a military solution in Afghanistan."[43] And in October, the US Senate Foreign Relations Committee rebuked the Bush administration for holding out for a military victory. As a State Department analyst put it, two changes had taken place since February 1989: "One, the congressional bipartisan consensus on Afghanistan is breaking up. And two, the perception that we are supporting a good cause is not there any more. We are no longer fighting the evil empire. They've gone. Now it's just Afghans fighting Afghans."[44]

Pérez de Cuéllar and the two officials now coordinating UN efforts on Afghanistan, Benon Sevan and Giandomenico Picco, saw an opening. In the fall of 1989, they attempted to launch a new UN-sponsored initiative to establish an intra-Afghan dialogue that would bring the opposition groups as well as the Kabul government to the negotiating table. Yet the United States, Pakistan, and Moscow all declined the initiative, which represented the main thrust of UN efforts in 1989. For the United States and Pakistan, any recognition of the Najibullah government's legitimacy continued to be anathema. Moscow continued to insist on an international conference, possibly fearing that the UN proposal might unite the chronically divided *mujahadeen*, depriving the Kabul government of a political advantage.[45]

Indeed, Moscow found it hard to let go of support for Najibullah. Although they may have expressed great frustration in private, Soviet leaders could hardly dump him, particularly at a point when he seemed to be gaining support within the country and increasingly capable of standing on his own two feet. In a private letter to Bush and again at the Malta summit in December 1989, Gorbachev insisted that that Najibullah could not be forced out prior to a settlement.[46] The two sides exchanged a number of accusations about their respective contributions to the problem and the failure to help find a solution. Shevardnadze even "sounded off" at US secretary of state James Baker, upset that the "friendly" relationship established between the two men during their prior meeting in Wyoming was not bringing concrete results. In a letter to Pérez de Cuéllar, Gorbachev noted that at Malta he and Bush had been able to focus their conversation on the need for a diplomatic solution. Yet he expressed his frustration that Bush refused to consider compromising on Najibullah: "It is important to see the realities of today's Afghanistan. It is necessary to take into account the fact that, following the withdrawal of Soviet troops, the government of the Republic of Afghanistan has felt more confident. We think that the opposition is starting to become convinced of this as well."[47]

At times, Moscow's support for Najibullah seemed to be as much a matter of decorum as defense of interests, at least as far as Gorbachev and even Shevardnadze were concerned. In December, the *New York Times* reported that although Shevardnadze and Gorbachev had both reiterated support for Najibullah during the Malta summit, they seemed to be dropping hints "with a wink and a nod" that he was dispensable.[48] Then, at a meeting with Baker in February 1990, a frustrated Shevardnadze reportedly blurted out: "Sometimes I wish all these people would just kill each other and end the whole thing." He went on to say that it would be better if Najibullah could stay at his post, but although "it would be very difficult for us to force him to go, it might be acceptable if he decided to leave on his own."[49]

US officials, for their part, were moving closer to the Soviet position on Najibullah. Although the *mujahadeen* showed no sign of dropping their insistence that Najibullah resign before negotiations could take place, the policy review and the changing atmosphere in Congress were moving the Bush administration toward dropping their insistence on such a scenario.[50] At the February 1990 meeting with Shevardnadze, Baker mentioned for the first time that the United States might stop insisting that Najibullah leave the scene before negotiations began.[51] Following the meeting, Soviet Foreign Ministry spokesman Gennadii Gerasimov said that although the formal US proposal for a settlement still "did not take into account the situation in Kabul and the solidity of the Najibullah government," the two sides had moved closer to a settlement based on intra-Afghan dialogue.[52] Najibullah announced publicly his desire for UN-monitored elections and his willingness to step down if defeated, and even suggested that he might resign before the vote was held during the negotiation process.[53]

A week after the meeting with Baker in February, Shevardnadze published a new set of proposals in the government newspaper *Izvestia*. His article repeated the call for an international conference and also called for a cease-fire, an end to US and Soviet arms shipments ("negative symmetry"), and elections monitored by the United Nations and the Islamic Conference Organization. Perhaps the biggest innovation was the idea that government and opposition forces could both hold on to the territory they controlled during the transition period.[54]

Following Shevardnadze's *Izvestia* article, Moscow and Washington seemed to move more quickly toward an agreement. At a meeting in Helsinki in March, US and Soviet experts elaborated on the proposals that came out of the Baker-Shevardnadze meeting in February and the *Izvestia* article. The failure of the Tanai rebellion confirmed that Najibullah still had enough support within the military and the party to hold on to power, even if it also highlighted the challenges he faced from rivals at the top. In May, US officials said that they would agree to Najibullah's participating in elections if he first stepped down. In testimony before the US Congress in June, Baker confirmed that "a very, very narrow difference" separated the views of Moscow and Washington.[55] President Bush, meeting with Pérez de Cuéllar in June, noted: "I was dead wrong about Najibullah—I thought he would fall when the Soviet troops withdrew." He went on to say that he could understand the Soviets' insistence on keeping Najibullah in office through the election period, and their insistence on following the "Nicaraguan model," where the Sandinistas remained in power while elections took place, then stepped down peacefully after the results were certified. President Bush's growing doubts about the situation in Afghanistan became evident again at a later stage in the discussion, when he asked Pérez de Cuéllar, "Is Hekmatyar a bad guy?" To which Pérez de Cuéllar responded: "I don't like him at all. He is a fundamentalist."[56]

Soviet leaders could take some comfort in the change in attitude and policy of other nations that had supported the *mujahadeen*. Relations with Iran had improved, as evidenced by Shevardnadze's high-profile visit to the country in February 1989, during which Ayatollah Khomeini hailed the Soviet withdrawal from Afghanistan.[57] In August, Kriuchkov and Shevardnadze noted that Iran was moving toward a more "constructive" position as a result of Soviet diplomatic efforts.[58] In October, Iran cut off military aid to Shiite insurgents, even encouraging them to work with the Kabul government.[59] From that point on, Soviet officials began to see Iran as generally playing a constructive role in Afghanistan. Similarly, China, which had once been part of the coalition supporting the *mujahadeen,* was no longer playing a hostile role.[60]

On the key question of US support for the *mujahadeen,* however, an agreement remained always just out of reach. Talks continued at the "expert" level, and Afghanistan was on the agenda at the July Baker-Shevardnadze meeting in Irkutsk and again when the two met in Houston that December. An international consensus, based on a draft prepared by UN officials, was within reach. Moscow now largely accepted the idea of a transition mechanism, "the powers of which could include important government functions."[61] Yet for all this progress, no agreement was reached in 1990. There were three reasons for this.

First, the Moscow leadership still found it hard to let go of Najibullah. The softening of the US position in the first half of 1990 and Najibullah's continuing hold on power encouraged Moscow to believe that sooner or later the United States (and Pakistan) would accept his involvement in a transitional government. Meeting with Najibullah in August 1990, Gorbachev reaffirmed his belief that the United States would ultimately recognize that they did not have someone better to offer as a national leader.[62]

Second, Moscow pointed out, with some justification, that even if the Soviet Union and the United States both cut off arms supplies, the opposition would still be able to count on support from Pakistan, Saudi Arabia, and other minor donors. Therefore, Soviet officials insisted on a more comprehensive settlement, even offering "negative symmetry plus" —that is, the withdrawal of heavy weapons (such as Scud missiles) from Afghanistan, if a complete cutoff of supplies to the *mujahadeen* could be guaranteed.[63]

Finally, internal political dynamics continued to play an important role, as they had in earlier periods. Although the internal debate in Moscow is difficult to trace for this period, there is reason to believe that Kriuchkov, and possibly Shevardnadze, continued to insist that Najibullah should not be forced to step down prior to the formation of a transitional government. It was only when both were finally out of government, in the fall of 1991, that Moscow agreed to a cutoff of arms supplies at the same time as the United States. Boris Pankin, the minister of foreign affairs who took over after the failed coup in August 1991, learned that the agreement had been prepared a year earlier, at the

Baker-Shevardnadze talks in Houston, but Kriuchkov had continued to block it.[64]

Handling the Home Front

Even before the withdrawal began, Gorbachev and other Soviet leaders had to decide how to explain the war to the public. The Soviet media had barely discussed Soviet military activities in Afghanistan prior to 1985; and in the years that followed, few details regarding either the origins of the war or its conduct emerged. As Andrei Grachev, an International Department official and later a Gorbachev aide, notes in his book *Gorbachev's Gamble,* there was serious concern among the leadership about the Soviet public's reaction to the withdrawal. While the public would probably accept the withdrawal, "official propaganda had been quite effective in concealing the truth about the real human price that had been paid."[65]

The withdrawal coincided with a real flowering of the Soviet media as a result of glasnost. The party no longer had absolute control over the press, and investigative journalism was emerging, allowing a wide range of investigative reporting on everything from the origins of the war to its conduct and aftermath. It was in the public sphere that Gorbachev scored the largest success for his Afghan policy. From 1988 through the collapse of the USSR, he managed to get all the credit for ending the involvement of Soviet soldiers in Afghanistan, without having to explain why it took him four years to do so. Gorbachev benefited from the genuine relief people felt that the "boys" had returned and no more would be sent to die; but his success also reflected a public-relations campaign that aimed to keep the focus on the origins of the war under Brezhnev.

Soviet leaders were already discussing how to explain the war and the withdrawal in the months prior to the signing of the Geneva Accords. At a meeting with several Politburo members working on propaganda and ideology, Aleksandr Iakovlev told his colleagues that press discussion of the war had to increase, but that avoiding all discussion of the war was a mistake: "God help us if we create the impression that our boys were put in harm's way for nothing, that they were needlessly disabled for the rest of their lives, that they fought for no reason. This is absolutely out of the question."[66] Soon after the accords were signed, the issue came up at a Politburo meeting. Gorbachev agreed with Iakovlev that propaganda should emphasize the "international duty" performed by Soviet soldiers, but at the same time it could not go too far: "After all, if everything was done right, why are we withdrawing?"[67]

Gorbachev proposed drafting a letter for the party and the country that would for the first time address the human and material costs of the war, as well as the reasons for the withdrawal. The letter had to strike a balance, "so that our withdrawal does not look like running away. [The letter] must stress that there is no military solution." The letter, circulated within the CPSU in May 1988, for the first time summarized many of the mistakes that had been made in the Afghan war. It spoke of the economic costs, the naïveté of Soviet party advisers, and the mistake made with the appointment of Babrak Karmal. Losses of men and matériel, long hidden from the party as well as the public, now received wide circulation:

> Combat action is combat action. Our losses in dead and wounded—and the CC CPSU believes it has no right to hide this—were growing, and becoming ever more serious. Altogether, by the beginning of May 1988, we lost 13,310 troops [dead] in Afghanistan; 35,478 Soviet officers and soldiers were wounded, many of whom became disabled; 301 are missing in action. There is a reason we say that each person is a unique world, and that when a person dies, this unique world disappears forever. The loss of every individual is very difficult and irreparable; it is difficult and sacred if one died carrying out one's duty.
>
> Afghan losses, naturally, were much heavier [than ours], including the losses among the civilian population.[68]

Although the letter avoided any actual mention that soldiers had died without good reason, it is hard to see how anyone reading it could avoid coming to precisely that conclusion. Indeed, time would show that Gorbachev and others were perfectly content to admit the war was a mistake, as long as this meant shifting the blame back to the Brezhnev era.

A month later, at the 19th Party Congress, Gorbachev echoed the sentiments of the letter to the assembled delegates. He pointed out that the Soviet leadership had to bear the "moral responsibility" for what had happened in Afghanistan, but quickly moved to distance himself from the mistakes of the Brezhnev era: "I must tell you that many Politburo members did not know about the decision [to send troops into Afghanistan]. I, for example, a candidate member of the Politburo, learned about the introduction of troops from the newspapers."[69] By January 1989, the Politburo was considering the possibility of publicly declaring that the decision to invade had been a mistake.[70]

In the new atmosphere of glasnost, the Soviet leadership could not hope to shape public perceptions of its policies through a monopoly on information. Although most of the leadership genuinely viewed the intervention as a mistake, they still worried about allowing criticism to get out of hand, thus undermining support for the military and other Soviet institutions.

As Gorbachev aide Vladimir Zagladin argued in a memo to his boss, thus far the Soviet press had continued to follow the old official line justifying the Soviet intervention in Afghanistan.[71] This line contained a number of contradictions. Among other things, Moscow had insisted that the intervention was necessary "to repel the foreign danger for Afghanistan." But at the end of December 1979, foreign support for the opposition was still minimal. "It assumed a serious scale only *after* our entry into Afghanistan and to a significant degree *as a result of the operation.*"[72] Zagladin concluded that while it was still too early to give a full explanation of what had happened, it was inadvisable to return to explanations that directly contradicted reality.[73] Zagladin accurately highlighted the contradictions, created by Soviet propaganda, that could potentially make trouble for the leadership on what was generally a popular move. In February 1989, in particular, it was far from clear if the Najibullah regime would last more than a few months. If Kabul fell soon after the withdrawal of Soviet troops and the decision to intervene had not been officially criticized, questions would almost certainly be raised about the advisability of withdrawing them, or perhaps about the conduct of the war under Gorbachev.

Zagladin's memorandum echoed concerns of some Politburo members, including Shevardnadze. Even as he was advocating a more hawkish line in support of Najibullah at the end of January 1989, the foreign minister pointed out that "within the party and in the country at large, there are different reactions to our withdrawal. At some point, we will have to announce that the introduction of troops was a gross blunder. . . . Many thought it was adventurism even then. But their opinion was not considered. Later the lies about successes started."[74] If years of Soviet propaganda were not reversed, Shevardnadze feared, the present leadership could be saddled with the blame for the outcome of the war.

In October, the new Congress of People's Deputies launched an investigation into the causes and consequences of the Soviet invasion. The investigative commission, headed by Georgii Arbatov, was given a mandate to interview military figures and officials involved in the initial invasion. Crucially, however, the commission did not focus on other aspects of the war.[75] Similarly, in October 1989, when Shevardnadze made a much-publicized admission to the Congress of People's Deputies that the Soviet invasion had "violated general human values," he moved to distance himself and his boss from responsibility for the war: "M. S. Gorbachev and I were candidate members of the Politburo. I found out about what had happened from radio and newspaper reports. A decision that had very serious consequences for our country was made behind the back of the party and the people. We were confronted with a fait accompli."[76] Several key military officers contributed, perhaps unwittingly, to Gorbachev's efforts to put some distance between himself and the emerging criticism of the war; they gave interviews in which they laid the blame on Brezhnev and his cohort.[77]

Even before October 25, 1989, when the Supreme Soviet officially condemned the decision to invade, a number of enterprising journalists, sometimes writing in newly established newspapers, began providing the public with previously unknown information about various aspects of the war, including the decision to invade. Relying primarily on interviews with participants as well as on their own experience covering the war in earlier years, these journalists for the first time presented completely new accounts of the war.[78] Official censorship rarely stepped in to block these exposés. And even when censors did move to block a piece from appearing in one media outlet, it could find its way to the public eye through another.[79]

Not surprisingly, journalists exposed not only the political mistakes made during the war, but also the brutal nature of the fighting, including the atrocities committed by Soviet troops. Issues like drug use and hazing *(dedovshchina)* were also being written about for the first time, with regard to service in Afghanistan as well as to army life in general. Needless to say, these sorts of investigative articles had been virtually unheard of in the Soviet Union. These revelations, and the emotional public reaction they evoked, contributed to the growing rift between the military and the civilian leadership, which often sided with the journalists.[80]

Similarly, the proliferation of groups like the Soldiers' Mothers Organization, which were at least partly a response to the war in Afghanistan, contributed to the loosening of state and party control over society. Some of these groups, particularly the veterans' organizations, had originally been formed within the framework of traditional party organizations like the Komsomol. By 1990, they were increasingly emerging as fully independent organizations, openly bypassing or defying state and party organizations. Their aims, however, had little to do with high politics and more to do with immediate concerns: medical and social aid for veterans, better treatment within the military, and so on.[81]

Crucially, neither Gorbachev nor Shevardnadze faced serious criticism for their policies on Afghanistan. Throughout the 1989–1991 period, criticism of the war focused on the decision to intervene and the management of the war in its early years. To some extent, this was a result of genuine curiosity on the part of journalists who wanted to understand the origins of the war. It was also the result of decisions by Politburo leaders to keep the focus of blame on Brezhnev and his circle.

Afghanistan and the Fall of the Soviet Union

When the withdrawal of Soviet troops was completed in February 1989, the USSR was still a superpower, with satellite states in Eastern Europe, allies throughout the Third World, and the Warsaw Pact organization all serving as seemingly unshakable counterweights to the United States and NATO. By the end of that year, most of the communist parties in Eastern Europe were out of power. A year later, East Germany was on its way to reunification with the Federal Republic of Germany, the Warsaw Pact had ceased to exist, and the USSR was facing economic catastrophe. By 1991, it was breaking apart at the seams.

In this context, it was probably inevitable that sooner or later Soviet material support to Afghanistan would cease or at least decrease to insignificant levels. Although Gorbachev still had at least the nominal allegiance of the military up to the time of his resignation at the end of December, the state was disintegrating.[82] Indeed, in 1991 the Soviet Union was able to deliver only 10 percent of the fuel contracted to the Republic of Afghanistan.[83] An Afghan delegate visiting the Soviet Union in the early fall of 1991 told a reporter: "We saw all these empty stores in Moscow and long queues for a loaf of bread, and we thought, 'What can the Russians give us?'"[84] Perhaps most damaging for Najibullah was the loss of political support that the Soviet disintegration entailed.

Najibullah harbored a justified fear that at some point or another he would be "abandoned." He was no doubt aware of the rumors circulating in the international press that Soviet officials were hinting, off the record, they might be willing to drop their insistence that he remain a player in a transition government. And as he watched the early stages of the USSR's ultimate dissolution, he expressed anxiety about where it might leave him. Najibullah was keenly aware that his support was in decline within Gorbachev's inner circle and among newly emerging political figures like Boris Yeltsin. In a New Year's greeting to Aleksandr Iakovlev, he reminded the "architect of perestroika" that the Afghans "would never forget those who helped our people in difficult, crucial periods of our history."[85] In August 1990, he complained to Gorbachev that "Yeltsin has publicly spoken in favor of ceasing aid to Afghanistan." He went on to defend Soviet assistance to Afghanistan, pointing out that while the USSR had carried out its obligations under the Geneva agreement, Pakistan and the United States had not. Gorbachev assured Najibullah that the USSR had no plans to abandon him; on the contrary, the United States was coming around to the Soviet view that the Kabul government had to be part of any transitional arrangement.[86]

Throughout 1991, Gorbachev's power declined. Republican leaders, particularly Russian president Boris Yeltsin, increasingly saw Gorbachev and the Soviet government as competitors for power. The governments of the various republics were declaring that their own laws superseded Soviet law; many were developing their own institutions, including ministries of foreign affairs, and working to get recognition abroad. Revenues were no longer reaching the Soviet treasury. Throughout the year, Gorbachev was preoccupied with establishing a new All-Union Treaty, and in July 1991 a draft was approved by the Supreme Soviet. It was due to be signed on August 20. On August 4, Gorbachev went on holiday; on August 19, a group of hard-liners, including Kriuchkov and Varennikov, tried to launch a coup. Their attempt failed, but it ended up taking Soviet power with it.[87]

As Gorbachev's power faded throughout 1991, his assurances to Najibullah rang increasingly hollow. When Kriuchkov and Shebarshin traveled to Kabul in April 1991, Najibullah asked for further confirmation of continued Soviet support and for more arms. Kriuchkov promised both. Later, Shebarshin told his boss that he doubted whether, in the developing political climate, such ongoing support was likely. Kriuchkov replied, rather sharply, that it would continue.[88] In April 1991, Kriuchkov was still in a position to give such promises. He was a member of the leadership and head of the KGB; furthermore, Gorbachev's "turn to the right" in the fall of 1990 and the spring of 1991 had brought him closer to Kriuchkov and other conservatives.[89] Yet much changed in the succeeding months. In June, Boris Yeltsin, who had advocated the cessation of supplies a year earlier, was elected president of Russia. The failure of the coup attempt led to the arrest of Kriuchkov and the ouster of Shebarshin.[90] Shevardnadze had already resigned at the end of 1990, embittered by attacks from conservatives who blamed him for giving up the Soviet empire.[91]

The fall of Khost, a provincial capital and a crucial strategic center, in April 1991 also began to change the calculus within Afghanistan. The rebels' takeover of that town weakened Najibullah and the Watan Party (the name, meaning "Homeland Party," used by the PDPA since 1990) much as the government's success at Jalalabad in March 1989 and the suppression of the Tanai coup in February 1990 had confirmed their ability to stay in power. Pakistan now began to harden its position, while the *mujahadeen* were eager to press their success further.[92]

The failure of the August coup in Moscow, whose plotters sought to arrest the USSR's disintegration by sidelining the reformers, had immediate repercussions for the Afghan problem. Only days before, Nikolai Kozyrev, who had been Moscow's chief negotiator at the Geneva talks and was still active in Afghan matters as an ambassador-at-large, had restated the Soviet position for "negative symmetry plus"—a guarantee that not only the United States but other parties like Saudi Arabia would cease supplying arms to the resistance.[93] Then on September 13, 1991, after a meeting between James Baker and the new Soviet foreign minister, Boris Pankin, the United States and the USSR signed an agreement to halt arms supplies to the belligerents and issued a statement confirming the right of the Afghan people to decide their own destiny without outside interference.[94] Kriuchkov's removal from the leadership proved crucial to the agreement that was finally being reached.[95]

In one sense, Kriuchkov's assurance to Najibullah held true. The Soviet Union never publicly renounced him, and indeed public avowals of support continued in the months leading up to its final dissolution. Although Moscow had been forced to abandon its insistence that arms supplies from Saudi Arabia and other sources be completely cut off, it did not have to give up Najibullah. Baker's statement after the signing ceremony implied that Najibullah was expected to stay on at least until the end of the election process, which would be organized by UN officials.[96] And Pankin writes that he had received private assurances from Baker that the United States would likewise press Saudi Arabia and Pakistan to stop supplying the *mujahadeen*.[97]

In the chaos of the Soviet Union's last months, rumors and whispers that Moscow might force Najibullah to resign—rumors that the press attributed to anonymous diplomats—became ever more common.[98] Afghan officials admitted that such reports were deeply demoralizing, but the Soviet government proved willing to reiterate its support as long as it was in a position to do so. In October, following a successful government defense of the city of Gardez against a rebel attack, Boris Pastoukhov, Soviet ambassador in Kabul, confirmed the old Soviet line that Najibullah's government could not be excluded from the peace process and declared support for the Watan Party leaders' proposal for a government of national unity.[99]

Such assurances counted for less with each passing week. Following the August coup, Gorbachev's standing within Russia and what was left of the USSR had fallen sharply, relative to Boris Yeltsin's. Yeltsin's voice increasingly dominated not only domestic issues but foreign-policy questions as well. Najibullah had expressed concern about this rising star in the summer of 1990. Now it was becoming clear that he needed to secure the support of Yeltsin and his associates before it was too late.

As the USSR was falling apart, Najibullah tried to make contacts with Russian leaders, perhaps sensing that there were now multiple centers of power in the Soviet Union which would be involved in deciding his fate. While Yeltsin initially seemed to react positively to the Afghan government's overtures, it soon became clear that his government did not believe Najibullah could hold on to power. In November, Yeltsin's vice president, Afghan veteran Aleksandr Rutskoi, met with a *mujahadeen* delegation in Moscow and told them that Yeltsin's government would "take all measures to bring peace to the long-suffering land of Afghanistan."[100] In the resulting communiqué, both sides expressed an understanding that "all power ought to be passed to an Islamic interim government."[101] Toward the end of 1991, Andrei Kozyrev and Aleksandr Rutskoi, who were in Pakistan to discuss the release of Soviet POWs, also made contacts with *mujahadeen* leaders.[102]

The division between the Soviet and Russian positions was also evident when representatives of both governments met with *mujahadeen* representatives in Moscow. Afghanistan, and the POW issue in particular, were becoming pawns in the political battle between Yeltsin and Gorbachev, and between the Soviet and Russian bureaucracies.[103] Rutskoi, apparently, let it be known that Russia would be willing to help depose Najibullah and accept the installation of an Islamic government. Rabbani seized on this and announced publicly that Russia was now finally ready to dump Najibullah, a fact Soviet officials immediately denied. Rutskoi, meanwhile, announced Russia's intention to cut off supplies of fuel.[104]

Lacking Soviet support and with no hope that Russia would help keep him in power, Najibullah knew that his days were numbered. His "lobby" in Moscow was out of the picture; the KGB, the institution which had backed him most forcefully, was in the process of being dismantled. Their importance, particularly in keeping Najibullah on life support in 1990–1991, was demonstrated by the rapid turn of events after August 1991. For Russian politicians, Afghanistan primarily represented a POW issue; they felt no sense of obligation to the Afghan regime, nor were they concerned about preserving their own superpower status.

When Bush and Gorbachev met at Malta in December 1989, the US leader expressed his disappointment with Soviet policies in the Third World. Bush pointed out that Soviet actions in the Third World were "out of step with 'New Thinking' and new Soviet directions in Eastern Europe and in arms control." Bush went on to say that "Soviet policies in regional conflicts were a major hindrance to the improvement of the overall US-Soviet relationship."[105] Indeed, Bush hit on a central paradox of foreign policy under Gorbachev: the New Thinking seemed to evolve much more slowly with regard to the Third World than it did in other areas of foreign policy, even though the Third World might be expected to be of lesser importance to Soviet prestige than, for example, Eastern Europe or arms control.

The Russian government that competed for power against the remnants of the Soviet regime in Moscow and finally took over at the end of

1991 did not feel bound by any commitment to the Kabul regime. Yeltsin had positioned himself as early as 1990 as an opponent of continued support for Najibullah, and no one in the Russian leader's circle seemed interested in pushing him in a different direction.[106] Gorbachev, on the other hand, was never able to break with Najibullah. Despite the rumors that frequently surfaced in the Western press about Moscow's willingness to stop supporting Najibullah, the Soviet government continued to push for his participation in a transition government until the very end. Unlike previous periods, however, Gorbachev was probably concerned more with the political than with the ideological ramifications of "abandoning" Najibullah. The Afghan leader still had influential backers in the Soviet government whose support was crucial for Gorbachev's political survival.

Gorbachev's priority in this period was maximizing the political gains of withdrawal while avoiding the fallout that might result if the Najibullah regime collapsed. The emerging free media consciously or unconsciously acquiesced in this. Most of the blame for the war was directed at the decision makers who had chosen intervention. Using Soviet forces to help the regime, even during a particularly difficult moment such as the battle of Jalalabad, was out of the question. By 1989, Gorbachev and his reformers had opened up the media and allowed the publication of articles quite critical of the war. There was no way to guarantee that even limited operations would not make him look like a hypocrite quite willing to continue using the Soviet military in "adventuristic" ways. Moreover, continued or renewed intervention could undermine his greatest foreign-policy achievements: his radical reorientation of relations with the United States, Europe, and China—a realignment that in 1989 still contributed to his popularity at home and made him a hero to many abroad.

Moscow did not have a coherent long-term policy for Afghanistan in February 1989, in part because many officials expected the Najibullah regime to collapse sooner rather than later. Gorbachev and many around him may have been hoping for at most a "decent interval" in those weeks after the withdrawal, a space of time prior to Kabul's defeat that would allow them to distance themselves from the war enough to minimize the political damage of Najibullah's defeat. The successful defense of Jalalabad in March proved that the regime could survive without Soviet troops as long as it had Soviet advisers and matériel. This was a boon politically, but it also meant that Moscow was still not rid of the Afghan problem and had to continue demonstrating its involvement and support. With Gorbachev increasingly distracted by the myriad domestic problems confronting him and uninvolved in Afghan issues on a daily basis, Kriuchkov could guarantee a basic level of support and fend off any suggestions about abandoning Najibullah and accepting US-Pakistani conditions for a transitional government.

On Afghan issues and foreign policy in general, in this period Gorbachev sought international consensus and agreement as a way to compensate for Moscow's rapidly declining ability to control events and negotiate from a position of strength. Soviet diplomats hoped that the United Nations would act to enforce the Geneva Accords, apply pressure on Pakistan and the United States, and in general take an active role in the formation of a new world order in which the USSR would be seen as a guarantor of peace. Yet UN officials proved reluctant to bear such a burden. Pérez de Cuéllar, for one, sought to avoid continued UN involvement. Even though the United Nations did continue to play a role (largely due to Soviet insistence), it proved capable of little beyond coordinating diplomatic efforts.

Finally, even as Gorbachev, Shevardnadze, and some of their advisers predicted (correctly) that the withdrawal from Afghanistan would help to improve relations with the West, they also hoped that this improved relationship would facilitate a solution in Afghanistan. They had sustained this hope since at least the fall of 1987. Gorbachev and Shevardnadze showed their frustration in meetings with American counterparts precisely because they believed that an understanding, albeit informal, was being breached each time the United States insisted on a settlement that excluded Najibullah and continued to provide support to the resistance. Their "cooperation" with the Bush administration in other areas was going completely unrewarded; by the end of 1990, they had acquiesced in the reunification of Germany and in the US-led operation against Saddam Hussein, a one-time Soviet ally. As with other questions of foreign policy, Gorbachev learned the hard way that his "friendship" with US leaders had some narrowly defined limits.

CONCLUSION

> We are leaving the country in a pitiable state. The cities and villages are ravaged. The economy is paralyzed. Hundreds of thousands of people have died.
>
> —USSR FOREIGN MINISTER EDUARD SHEVARDNADZE, January 1989

In January 1992, the war in Afghanistan had already entered its thirteenth year and showed few signs of ending. The Watan government still held Kabul and most provincial capitals; but with no more Soviet aid forthcoming (and no Russian aid to replace it), it could not hope to fight the *mujahadeen* indefinitely. Najibullah was economically and politically isolated. Although his military still had plenty of Soviet planes, tanks, and weapons with which to carry on fighting, the elimination of fuel supplies was taking its toll. Najibullah's air force, which provided a crucial advantage over *mujahadeen* forces, was grounded. The government was forced to spend its rapidly depleting currency reserves on fuel from Iran.[1]

Najibullah's last months saw the unraveling of his network of patronage and of the militias that had kept his regime in power since the Soviet withdrawal. Lack of funds was one of the reasons; the other was an ill-advised attempt to preserve some influence by once again appealing to Pashtun solidarity. Toward the end of January 1992, Najibullah tried to replace General Abdul Mumin, the Tajik commander of the Hairaton garrison, with an ethnic Pashtun, General Rasul, a former commander of the Pul-i-Charkhi prison during the Khalqi period. When Mumin refused to give up his post, he found a willing ally in General Rashid Dostum, the Uzbek militia commander and Ismaili leader previously loyal to Najibullah. Mumin also reached out to Massoud, telling him he had plenty of weapons but not many men. Abdul Razak, a young engineer and graduate of the Kabul Polytechnic working with Massoud, later recalled: "At the time, I was working on building a small hydroelectric generator. I asked to go with the fighters to Mumin. We went on foot for three days, and on the third day we met Mumin on the banks of the Amu Darya. We ate and we talked. Mumin said to us: 'Don't think you've won. It's we are who are helping you!' We didn't mind. We spend three days there, and they gave us machine guns and pistols. And then we went toward Mazar-i-Sharif."[2] By the end of March 1992, a coalition that included Massoud, Dostum, and a number of other Uzbek and Ismaili commanders controlled the northern provinces and stood poised to take Kabul.[3]

One by one, Najibullah's remaining allies abandoned him. By April, Kabul was surrounded by Massoud's forces closing in from the north and Hekmatyar's advancing from the south, in a preview of the carnage that would envelop the country in the years to come. UN officials continued to work for the creation of an interim government; but as had often happened previously, their efforts were overtaken by events. On April 12, Najibullah called the most senior of the seven remaining Soviet (now Russian) officer-advisers in Kabul to his residence. Power would soon be in the hands of the opposition, Najibullah said, and it was time for the officers to leave. The Soviets were traitors, he said, but nevertheless he felt obligated to see that they were sent home safely. The next day, Najibullah met the seven officers at the airport, personally making sure that the plane took off without incident.[4]

A day later, on April 14, Najibullah was confronted with the presence of a militia led by Rashid Dostum within the capital. Dostum's loyalty had been crucial over the past few years, particularly in putting down the Tanai coup in 1990. Now Dostum, sensing that regime's collapse was imminent, began acting as a free agent, looking to ally himself with the forces that would soon take Kabul. Without him, Najibullah's government, or what was left of it, could not hope for even a decent bargaining position vis-à-vis the forces threatening to take the city.

In 1989, Najibullah had rejected Soviet offers to take refuge in Moscow. By April 1992, he understood there was little hope of holding on to power or even making a graceful exit. The president of Afghanistan went to the UN compound in Kabul and asked for help to leave Afghanistan and join his family in India. When he arrived at the airport several days later, accompanied by his bodyguards and several UN officials, he found it surrounded by Dostum's militia. His escape blocked, Najibullah spent the next four years living in the UN compound.[5]

Russia played a minimal role in these final months. Although Russian diplomats were involved in trying to secure Najibullah's safe passage, Moscow in no way sought to continue playing an influential role in Kabul. The Russian government's attitude was summed up in a statement by Kozyrev: "Everything in Afghanistan is ready for settlement. The only problem is the Soviet support of 'extremists' led by Najibullah."[6] As Evgeni Ostrovenko, the ambassador of the Russian Federation sent to Kabul in 1992, told an interviewer: "By early 1992, the regime had outlived its time. We Russians had nothing to do with it."[7]

A number of former Soviet participants later spoke out against the Russian government's handling of the Afghan crisis. Varennikov, Kriuchkov, Egorychev, and others have pointed to the "betrayal" of Najibullah by Russian leaders as the reason for the chaos that later enveloped Afghanistan.[8] Indeed, it does seem that Yeltsin's foreign-policy team used the Afghanistan issue to distance themselves from Gorbachev and to identify him with the more notorious aspects of the Soviet regime, while elevating their own status as true democrats and reformers. It may also have been a way for Yeltsin to raise his profile among foreign leaders, who only grudgingly began to accept him as the leading figure in Moscow.[9]

In any event, the economic basket case that was Russia in January 1992 could have done little to support Najibullah. The military was on the verge of collapse, and the Central Asian republics were no longer reliable staging grounds for any kind of support to Kabul, making the logistics of any such operation very difficult. Most important, Russia had very little diplomatic clout. Dependent on foreign aid to feed its citizens and on institutions like the International Monetary Fund and the World Bank to prop up its collapsing economy, it could have done little to help Najibullah even if it had wanted to.

For the United States, too, the Afghan adventure was over, at least for the time being. Some of the politicians who helped to fuel the anti-Soviet *jihad* were either leaving the scene or quickly growing disillusioned. Even Charlie Wilson, the iconic booster of the "freedom fighters" in Congress, found it hard to muster much enthusiasm for events in Afghanistan. The *mujahadeen* were now dealing in drugs and using US-supplied weapons to kill each other; the slaughter of Afghan civilians continued. The Clinton administration, which came into office in 1992, had other priorities. (Afghanistan had not been on the foreign-policy platform of any major candidate in the 1991 election.) Once in office, Bill Clinton wanted to focus on Africa and showed little interest in Afghanistan. Although some in the CIA and the State Department argued that the policy of letting Afghans fight among themselves would ultimately backfire, both agencies turned their resources to other areas.[10] Charles Cogan, one of the key protagonists in the war on the CIA side, wrote in *World Policy Journal* that the United States "was no more able to put together a polity in the ghost town that Kabul had become than it can in Dushanbe or, alas, in Mogadishu. Nor should it try."[11]

The United States was also growing apart from its main partner in the Afghan fight, Pakistan. The Afghan war had saved the US-Pakistani relationship after Zia's coup; but with that common cause out of the way, old problems resurfaced. Pakistan was working on a nuclear weapon, to which US officials had turned a blind eye for years. During the first meeting between Benazir Bhutto and George H. W. Bush, the US president explained that with the Cold War over, it would be very difficult to avoid invoking the Pressler amendment, which stipulated that US aid was contingent on Pakistan's giving up its nuclear weapons. After October 1990, US aid to Pakistan was frozen. Pakistan's support for insurgents in Kashmir further inflamed tensions with the United States—tension which carried over into the Clinton administration. Without the Afghan war, Pakistan lost influence in Congress, while supporters of India became more vocal.[12]

The end of Russian/Soviet and US involvement in Afghanistan did not bring peace to that unfortunate country. The *mujahadeen* proved that opposition unites much better than power does. By mid-April, Massoud had control of Kabul airport, seized initially by Parchamist troops loyal to Karmal's brother, Mahmud Baryalai. Massoud avoided entering Kabul (despite the requests of some Parchamists), probably because he was afraid of sparking a confrontation with Hekmatyar. Instead, he urged the alliance leaders in Peshawar to come to an agreement on an interim government. Hekmatyar, meanwhile, closed in on Kabul from the south and made contacts with sympathetic Khalqi Pashtuns, who allowed him to infiltrate fighters into the city and provided them with arms. On April 25, Massoud and his allies finally entered the city and forced out the Khalqi/Hizb-e fighters. Order in the capital broke down as armed groups roamed the city in an orgy of looting and destruction. Having survived the past thirteen years relatively unscathed, Kabul now felt the full brunt of war.[13]

Although the alliance leaders did eventually come to an agreement on an interim government, they failed to gain either legitimacy or real control in much of the country. Instead, the country devolved into an amalgam of territories controlled by warlords. Some areas, like Herat, which fell with little fighting to the warlord Ismail Khan, gained a measure of stability and even economic revival. The Hazaras were able to maintain autonomy in their own territory, and Rashid Dostum, with Uzbekistan's support, controlled his own fiefdom. Massoud controlled most of the northeast, while Helmand in the south was contested by several militias but was increasingly the fiefdom of drug lords. Kabul itself was the scene of continued fighting, and a stream of refugees left the city.[14]

By this point, Dostum had parted ways with the Rabbani/Massoud forces controlling Kabul and was attacking the capital sporadically, adding to the destruction wreaked by Hekmatyar. Kandahar, Afghanistan's second-largest city, was even more chaotic. As warlords competed for power, they seized farms, homes, and businesses and divided those among supporters. The road from Kandahar to Quetta, a crucial link for Kandahar's economy, was controlled by numerous armed bands, each of which demanded payment from truckers and other travelers.

This was the context in which the Taliban movement emerged in 1994. Led by veterans of the anti-Soviet and anti-Najibullah *jihad* after 1989, these students and sometime fighters united to disarm the population, establish a strict rule of law, and reunite the country. With the mullah-warrior Muhammed Omar taking the lead, they began by intervening in and settling local disputes, and freeing captured girls and boys who were being held to pleasure local commanders. By the end of 1994, they had routed Hekmatyar's forces on the Pakistani border, cleared the rode to Kandahar, and taken the city itself. (Pakistan had by this point grown disillusioned with Hekmatyar, and the ISI was beginning to look at the Taliban as the best chance for preserving Pakistani interests in Afghanistan.) From there, the Taliban army grew, incorporating some groups of fighters—many of whom surrendered, rather than take on the increasingly mythologized and seemingly unstoppable force—and decimating others. Everywhere they went they brought their uncompromising and reductionist interpretation of Sharia law, which included closing down girls' schools, forcing all men to grow beards, forbidding television, radio, and sports, and imposing harsh penalties for transgressions. Thieves lost hands and feet; women found guilty of disobeying the Taliban's injunctions were stoned.[15]

Although the movement's leaders had taken part in the *jihad* of the 1980s, many of the footsoldiers had no first-hand knowledge about it. Rather, they were boys raised in madrasas (Islamic religious schools), often children of parents who been killed. They were completely unmoored from Afghan society at large, and the utopian future they fought for was imagined for them by their commanders and mullahs. The journalist Ahmed Rashid wrote: "These boys were what the war had thrown up, like the sea's surrender on the beach of history. They had no memories of the past, no plans for the future, while the present was everything. They were literally the orphans of the war, the rootless and the restless, the jobless and the economically deprived, with little self-knowledge."[16] In 1996, these "boys" took Kabul. Among their first acts after taking the city was to seize Najibullah from the UN compound where he had lived for the previous four years; they killed him and hung his body from a lamppost. Massoud would later claim that he had tried to persuade Najibullah to come with him as his forces retreated from Kabul, but the proud former president had refused, believing he was safer with the approaching Pashtuns.[17]

The Taliban earned Pakistan's support because they held the promise of restoring order and of being useful to the ISI. Similarly, the United States largely turned a blind eye to the Taliban and their excesses. Partly this was a result of the Clinton administration's turn away from Afghanistan. Even as the Taliban's cruelty became widely known, US officials probably breathed quiet sighs of relief that at least some order seemed to be returning to Afghanistan. They also hoped that with Afghanistan united, the US oil company UNOCAL would have a chance to build a proposed pipeline that would run from the Caspian Sea to the Indian Ocean. In 1996, UNOCAL representatives and US officials met with Taliban leaders to hammer out a deal. They never came to an agreement.

US engagement in the next few years was limited to tepid efforts to encourage talks between the Taliban and the increasingly fragile Northern Alliance forces still fighting them. Despite the Taliban's violence and repression within Afghanistan, the only major concerns for US officials at the time were the regime's harboring of Osama bin Laden, another former anti-Soviet *mujahadeen* of Saudi origin who was responsible for several major terrorist attacks on US targets, and the Taliban's involvement in the opium trade. From late 1999, the Taliban were subject to UN sanctions, imposed with US support. Still, the idea that was floated by some CIA officials to start supporting the Northern Alliance—forces that were coalescing around Ahmad Shah Massoud and that were opposed to the Taliban—gained little traction.[18]

Ironically, one of Massoud's few sources of support in this period was Moscow. When civil war broke out in Tajikistan in 1992, the Russian 201st Division spearheaded a Commonwealth of Independent States (CIS) peacekeeping force, and Russian guards were deployed along the border with Afghanistan. Although Massoud and Moscow were backing different groups, ultimately both were pushing their clients toward a compromise solution. Massoud played the role of mediator during that conflict, helping to forge the deal which allowed the pro-Russian president to retain power. In 1998, with the Taliban pressing on former Soviet borders, Massoud sent his minister of foreign affairs to Moscow, where he met with Yeltsin's representatives. Russia became one of Massoud's sponsors, along with Uzbekistan and Tajikistan.[19] Massoud's earlier contacts with Soviet officers and diplomats helped to win him a sympathetic ear in Moscow. But the Northern Alliance would not get serious foreign aid until after the attacks on the Pentagon and New York's World Trade Center on September 11, 2001. By then, Massoud was out of the picture: on September 9, he was killed by assassins posing as journalists.

Unlike the Soviet Union, the Russian Federation in the 1990s did not see itself as the liberator of Third World states. Indeed, many Russians felt that such aid had helped to undermine and impoverish their own country. The Russian Federation, so far, has not intervened militarily in support of any foreign government or movement, aside from minor engagements in the CIS.[20] Only in the past five years has Moscow, buoyed by high energy prices, been able to play a serious role abroad. During the 1990s, its military efforts were limited to trying to arrest the process of disintegration that had led to the breakup of the Soviet Union.

Soviet and US interventions in the Third World played an important part in shaping the Cold War and the history of the newly liberated states emerging from the collapsing European empires. For Soviet leaders, the success or failure of their clients in these Third World states often had both ideological and strategic significance: ideological because success proved the superiority of their model of modernization; strategic because success helped to maintain the balance of power in the world and to prevent US domination. In 1962, the desire to protect the Cuban revolution and also to balance Washington's superiority in intercontinental nuclear missiles had led Khrushchev to install Soviet atomic weaponry on Cuba, taking the world to the brink of nuclear war. In later years, Moscow's aid and interventions had similar dual motivations. The recent historiography on Soviet intervention in Hungary (1956) and Czechoslovakia (1968), and the role of military aid and advisers in the Horn of Africa and the Middle East, confirm this.[21]

Soviet involvement in the Third World, particularly direct involvement, always had its critics within the CPSU and the government. These critics argued that Soviet involvement in the Third World brought few benefits to the USSR, while at the same time undermining détente with the United States. Throughout the 1970s, they were either overruled or ignored altogether. In March 1979, as panicked Afghan communists asked for Soviet military power to put down a major uprising in Herat, the critics' views held sway, preventing an intervention. By December of that year, their ability to influence decision making had been eroded by the more persistent lobbying of those who saw intervention as the only way to protect Soviet strategic interests and prestige.

Yet the decision to send troops to Afghanistan also represented the apex of Soviet interventionism and indeed of Soviet involvement in the Third World. Russian disillusionment with involvement abroad in the 1990s had its roots first and foremost in that fateful decision made at Brezhnev's dacha in 1979. As the extent of the quagmire became evident, Soviet leaders, even the arch-interventionists, began to reconsider the value of propping up friendly regimes with Soviet troops. When protests threatening the socialist government in Warsaw erupted in 1980, Andropov, a key figure in the decision to intervene in Hungary, Czechoslovakia, and Afghanistan, rejected this option for the Polish crisis, saying, "The quota of Soviet interventions abroad has been exhausted."[22] The so-called Sinatra Doctrine, which allowed socialist regimes in Eastern Europe to collapse in 1989, rightly belongs to the Gorbachev era; but its roots lay in the early 1980s, when Soviet leaders began to feel the full effect of the hangover that resulted from their overindulgence during the previous decade.[23]

Thus, not only had Soviet leaders *before* Gorbachev already decided that the Soviet intervention in Afghanistan was too costly and had started looking for a way out; they had also begun reevaluating the value of Soviet involvements more generally. Afghanistan, of course, was quite different from any other Soviet involvement in the Third World. The stakes were simply higher. For one thing, there was the sheer scale of the intervention. True, Soviet advisers and even pilots had helped armies in Africa, Latin America, and Asia fight their enemies, but here, for the first time, Soviet troops were involved en masse, essentially taking on the primary duties of the host country's military. In this way, the USSR's intervention in Afghanistan resembled those in Hungary and Czechoslovakia much more than its involvement in the Arab-Israeli conflict or in any of the localized conflicts in Africa. This made the intervention more costly, and it raised the stakes: the loss of the Kabul government could mean not only an ideological defeat, but also a military one—something which both Gorbachev and his predecessors sought to avoid at all costs.

In Afghanistan, also, Moscow was confronted with a popular uprising against a client government's rule—a rebellion which it had not faced elsewhere. The resistance in Hungary and Czechoslovakia, which did shake Soviet leaders in 1956 and 1968, barely merits comparison with the much wider *jihad* against the Kabul government and the 40th Army. In those cases, protesters in the capitals were quickly dealt with by overwhelmingly superior force and were labeled counter-revolutionaries. The countryside stayed quiet. In Afghanistan, Soviet leaders saw clearly that the resistance had popular support and that the Kabul government had few friends in the countryside. Prior to the intervention, many in the Politburo realized that Soviet troops, if they were sent in, would end up fighting common Afghans. From the ideological perspective, this would be a disastrous situation that could be easily exploited by the USSR's enemies—a tricky issue which no amount of counter-propaganda could undo. Yet rather than impelling Soviet leaders to beat a hasty retreat, this situation also raised the stakes and moved them to look for victory. If the Kabul government could somehow be made more palatable to the population, then a Soviet intervention, perhaps, would not look like a police action directed against the peasantry of a poor neighboring country.

Finally, as Soviet diplomats never tired of pointing out in negotiations, the Soviet Union and Afghanistan shared a 2,000-kilometer border. Throughout the war, Soviet leaders did not worry much about a spillover effect—the possibility that the war would ignite serious uprisings in Central Asia. The issue was different: Afghanistan could become yet another state used by the United States and NATO to surround the USSR. As other historians have shown, and as I noted in the Introduction, the fear that Afghanistan would become a base for missiles directed at Moscow was one of the key factors motivating Andropov and Ustinov to push for intervention. Many years later, Kriuchkov and Shevardnadze made similar arguments as they urged Gorbachev to do everything possible to protect the Najibullah regime from collapsing. Indeed, as Shevardnadze argued, the Soviet campaign in Afghanistan meant that "anti-Sovietism" would exist there for a long time—and the country would now be more open to taking an anti-Soviet stance and making itself available for the USSR's enemies.[24]

High as the stakes were, Moscow was not prepared to expand the war. Although one of the 40th Army's main strategic problems was cutting off the supply of arms coming from Pakistan, Soviet leaders never took serious punitive measures, either military (bombing the training and supply camps on the Pakistani side of the border) or diplomatic (breaking diplomatic relations). Such an expansion of the war, on the model of the US bombing of Cambodia and Laos during the Vietnam war, would have made sense militarily but would have caused further isolation for Moscow. The refusal to expand the war shows that Soviet leaders were trying to minimize the extent of confrontation caused by the invasion.

It also points to a certain degree of confidence among Moscow's leaders that their political advice and technical assistance could help overcome the numerous difficulties faced by the PDPA regime. Parallel to the military effort of the 40th Army (and Soviet advisers in the armed force of the RA), there was also a smaller army of political advisers, technicians, educators, and similar personnel that undertook a modernization and nation-building project in Afghanistan. These advisers and specialists dug ditches, operated mines, extracted natural gas, wrote speeches on behalf of politicians and memoranda on behalf of ministers, and went out into the countryside to help Afghan communists reach out to the local population. The last of these efforts may have often done more harm than good, but the technical advisers, at least, did create tangible benefits for many Afghans: factories provided employment, medical clinics brought modern health services to areas where they were previously unheard-of, and extraction of natural resources helped to keep the government solvent throughout most of the 1980s.[25] Needless to say, these benefits were poor compensation, in the eyes of ordinary Afghans, for the carnage wrought by revolution and war.

Soviet leaders believed that they needed to undertake a nation-building project in order to stabilize the country and bring their troops home. Moscow had been supplying technical and political advisers since the 1950s, and sent even more after the Saur Revolution. It was the invasion, however, that turned this assistance into a nationwide project. Building socialism was not the goal. Soviet leaders believed that the country was not ripe for socialism and urged their tutees in the PDPA to move away from a revolutionary agenda.[26] The goal was political stabilization, with *modernization* as its major tool. That this modernization often looked like socialism stemmed from two factors. First, the PDPA leaders thought of themselves as revolutionary Marxists and shed this coat only reluctantly; second, the advisers sent by Moscow, particularly the party and agricultural advisers, knew only how to replicate their experience in the USSR and therefore could not (or would not) shed the their ideas about what modernity was.

The Soviet experience in Afghanistan was thus a culmination of the USSR's other Third World involvements during the Cold War. For decades the Soviet Union had been offering a version of modernization, sending its military, political, and technical advisers to emerging states that were socialist or leaning that way. Soviet modernization was a challenge to colonialism and to the American model—although, as practiced in the context of counterinsurgency warfare (for example, in Vietnam), the two models looked remarkably similar.[27] Since the scale of the effort in Afghanistan was so grand, the potential for failure was considerably heightened as well. The power and influence of the USSR rested on several pillars: its military might, its technological prowess, and the superiority of its political model for achieving modernization and fending off neocolonialism. It cannot be ignored that the specter of a high-profile failure, the kind that might reveal the vulnerability of all three of these pillars, hung over Soviet leaders as they tried to plot a course out of the Afghan quagmire.

A fourth pillar of Soviet power was loyalty to friends, something the country's leaders believed in quite firmly. There may have been some abstract notions of honor involved, but there were also geo-strategic and ideological reasons. A state that abandoned its allies in difficult times would not hold on to its global influence for long. Individuals (like

Karmal) could be expendable, but entire governments certainly were not. It is hardly surprising that the imperative of "not abandoning Najibullah," and the potential reaction among other Soviet allies in the Third World if Moscow were to do so, came up repeatedly in Politburo debates. Of course, the Soviet leaders were not alone in thinking this way. Such thinking was typical of superpower politics, and echoes of similar concerns could be heard in the US debate about South Vietnam and Taiwan.

From 1982, when Moscow began to look seriously for a way out of Afghanistan, to 1989, when the withdrawal of troops was completed, Soviet leaders worked to buy time for their modernization and political strategies to work. They may have believed their generals when the latter said that there was no military solution to the Afghan problem, but they still needed the 40th Army to provide the breathing space for their other strategies to work. Those strategies included the modernization program discussed above, the effort at National Reconciliation (both after the initial invasion and in its reincarnation in 1987), and the shuffling of leaders at the top. The Soviet strategy also included the diplomatic effort, undertaken through the United Nations and other channels, to secure recognition for the government in Kabul and at the same time create a legal framework for a Soviet withdrawal. Important as that effort was, however, changes in US-Soviet relations were more consequential.

Although Gorbachev understood the importance of bringing Soviet troops home in early 1985, the imperative of protecting Soviet prestige and relations with client states, as well as avoiding the domestic "ideological damage" of a failure in Afghanistan, led him to support a series of initiatives during his first three years in power. These measures included the replacement of Karmal with Najibullah, the launching of National Reconciliation, the resuscitation of the Geneva talks, and a diplomatic push on all fronts.

The decisive turn that Afghan policy took in late 1987 and early 1988 was motivated by two factors. The first was Gorbachev's realization that none of these initiatives had done or would do much to stabilize Afghanistan. The Policy of National Reconciliation had stalled because of resistance from within the PDPA and a lackluster response from opposition forces. Najibullah proved a more capable leader than Karmal, but he was no panacea. Talks with Najibullah in July 1987 had left Gorbachev deeply disappointed.[28] By mid-1987, the consensus in the Politburo on what could be achieved in Afghanistan had changed dramatically. Most of the leadership was willing to accept a secondary role for Moscow's client within a future Afghan government.[29]

The diplomatic effort to find a settlement on Afghanistan followed the contours of the US-Soviet relationship in the 1980s. In mid-1987, Moscow's Afghan policy seemed to be failing, but US-Soviet relations were improving. The second half of 1987 became a major turning point in the relationship between the countries. After two important but ultimately unsuccessful summits (Reykjavik in 1986 and Geneva in 1985), the groundwork had been laid for a summit in Washington and then in Moscow. Gorbachev and his foreign-policy team knew that a resolution in Afghanistan would go a long way toward improving relations with the United States. At the same time, Gorbachev came to believe that an improving relationship with the United States could help to secure the kind of settlement in Afghanistan that he and his predecessors had been looking for.[30]

The changes in Moscow's approach to the Afghan problem in 1987 and 1988 were related to broader changes in the Politburo's approach to foreign-policy problems. This was the period when, thanks to Gorbachev's efforts, like-minded reformers had been brought into the Politburo and conservative politicians, including Andrei Gromyko, were pushed out. The slow pace of change in Gorbachev's first two years pushed him to try for more radical approaches, and he increasingly linked the success of his foreign policy to improved relations with the West.[31] This served as an added incentive to find a way out of Afghanistan, even at the risk of abandoning the key principles that had kept Brezhnev, Andropov, and even Gorbachev himself from reversing the intervention. The period 1987–1988 saw the most profound change in Moscow's Afghan policy since the intervention. Yet it was not completely irreversible. Once the withdrawal had actually begun and it looked like the long-feared collapse in Kabul might actually take place, Gorbachev reverted to supporting more aggressive policies, even at the risk of aggravating tensions with the West.

One scholar, who wrote about the Soviet withdrawal from Afghanistan in the mid-1990s, has argued that it represented the triumph of reformist thinking over conservative elements in Moscow and identifies this change primarily with Gorbachev.[32] While it is true that this reformist thinking contributed to the change in how Soviet leaders formulated foreign policy and viewed their commitments and rivalries, the story of the withdrawal reveals remarkable continuities as well. With the minor exception of Gorbachev's first years in power, Afghan policy was made in Moscow by a small group of men, who often shut detractors out of the decision-making process. During the first years of the war in particular, when Afghan policy was dominated by the heavyweights Andropov, Ustinov, and Gromyko, outmaneuvering them proved almost impossible. Policy could be altered only when these leaders had changed their minds, which started to happen in 1981. During the Gorbachev years, it was difficult but not impossible to outmaneuver Kriuchkov and Shevardnadze, who had come to dominate Afghan policy, though such a move could carry serious consequences.[33]

This last point raises another question: How did those who came to dominate decision making on Afghanistan reach their individual conclusions about what policy should be followed in that country? This book has shown how decisions were made within the Politburo as a whole and within the broader mindset of leaders who made Afghan policy. Lack of solid evidence makes it difficult to answer this more specific question definitively; but on the basis of material presented here, several factors need to be noted.

The first factor is the impact of reporting from junior officials working in Afghanistan. Unjustifiably positive reporting was a problem in many areas of Soviet bureaucracy, and it almost certainly contributed to Soviet leaders' misunderstanding of the situation in Afghanistan. Advisers and other Soviet officials working in Kabul had every incentive to make their reports more positive, since these reports were evaluations of their own success. True, there were some negative, critical reports—but if leaders like Andropov, Gromyko, and Ustinov did not shut such reports out completely, at least in the first year of the war they seemed to balance negative assessments of their Afghan policy against the positive ones. Thus, it is not surprising that a similar refrain could be heard in Politburo discussions of the Afghan problem from 1980 until 1987: there were still problems, but progress was being made, so the right thing to do was to extend the Soviet presence in Afghanistan until the problems were solved. The debates after Gorbachev's realignment of Soviet Afghan policy showed that no one was immune to this line of thinking. Kriuchkov and Shevardnadze fell into the same trap: they believed that Najibullah was making progress toward forming a more stable government, and they used this as an argument against curtailing Soviet support for him.

A second factor, related to the first, was the rivalry between different agencies working in Afghanistan. The rivalry affected decision making, because whichever agency had the most effective sponsor in Moscow would have the edge in presenting its point of view. In these turf wars, the military usually ended up the loser and the KGB the winner, as happened in the fall of 1988. Indeed, Soviet officials were often acting as proxies in the intra-PDPA power struggle, championing the position of their advisees in Moscow. As a result, decisions in Moscow sometimes reflected a preference for an Afghan faction or leader, even if that faction or leader did not necessarily act in Moscow's broader interests.

A final factor—the most subjective in nature and hence the most difficult to evaluate—consisted of the internal politics and power struggles within the Politburo. Andropov, Gromyko, and Ustinov were all potential successors to Brezhnev, and there were other contenders as well. Any major failure in Afghanistan would reflect poorly on them and jeopardize their chances of successfully assuming a high post, or the top post, in a post-Brezhnev government. Indeed, a post-Brezhnev leadership might well look for scapegoats if faced with a disaster in Afghanistan. For these men, Afghanistan had become a test of their resolve—their ability to cope with setbacks, to defend Moscow's allies, and to see a foreign-policy crisis through to a satisfactory conclusion. Only when they were convinced that their military, economic, and political efforts within Afghanistan were not going to bring the desired result did they turn to UN diplomacy and reconsider other channels they had rejected earlier. Gorbachev and Shevardnadze shared similar concerns: they were reformers with little foreign-policy experience, and a disaster in Kabul could call into question their ability to guide the Soviet Union through a difficult crisis.

The recent US-led interventions in Afghanistan and Iraq have prompted scholars, officers, and policy makers to reexamine the nature of counterinsurgency at the tactical and political levels.[34] The dilemma facing Soviet leaders was similar to that of other politicians managing a counterinsurgency. The most useful comparison in the Cold War context is, of course, the US counterinsurgency campaign in Vietnam. There are enormous differences between the two cases—not least, the very different kind of domestic pressures US and Soviet leaders faced. But there are some very useful parallels to consider.[35]

In both cases, the military's involvement was only part of the picture.[36] In the early years of the Vietnam war, US policy makers, inspired by modernization theory, undertook initiatives like the Strategic Hamlets Program to win over Vietnamese peasants and show that the government of Ngo Dinh Diem could provide the peasants with economic and military security. Both the US effort in Vietnam and the Soviet effort in Afghanistan were motivated by a belief in a type of modernization, an ideology that often did not work well with the prevailing social and political realities. In addition, as with Soviet "clear-and-hold" efforts, US advisers played a key role not only in military operations but also in political and development efforts. The Strategic Hamlets Program, abandoned in 1963, resurfaced under President Lyndon Johnson as the "New Life" Hamlets Program. Ultimately, however, US leaders relied on military power and international diplomacy to bring the troops home.

Although the wars in which the superpowers were involved were unpopular at home and among the international community, in both cases their elites were concerned about the way that withdrawal would impact their countries' credibility, and whether it might not lead to the collapse of other allies under pressure from insurgent movements. Soviet leaders feared undermining the USSR's position and authority as the leader of the communist world and supporter of national-liberation movements. The possible effect of a defeat in Afghanistan on the Soviet Union's reputation was a concern not only of Old Thinkers like Leonid Brezhnev and Yurii Andropov, but also of the reformist group that dominated the Politburo after 1985, which included Gorbachev himself. In the case of the United States and Vietnam as well, there was continuity in this regard, stretching from the administration of Dwight Eisenhower to that of Richard Nixon.

Crucially, the presence of the United States and the USSR in Vietnam and Afghanistan, respectively, was extended by the elites' belief in what their country could accomplish through military might, alongside the aforementioned modernization programs and political advice. Thus, even though Soviet leaders recognized that the Soviet example was inappropriate for Afghanistan to follow, they believed they could go a long way toward stabilizing their client government in Kabul through a mixture of political tutelage and modernization programs. Soviet leaders prolonged their country's presence in Afghanistan because of a desire to give their programs there a chance to work, much as US leaders continued to believe that their military victories and various initiatives within Vietnam would bring about desired results, long after it became clear that overwhelming superiority of technology, military capability, and resources were not bringing success.

Similarly, in both cases the clients proved capable of manipulating the patrons to their own advantage, thereby extending the superpowers' involvement. Despite a general consensus at the top of the Soviet hierarchy on Moscow's goals in Afghanistan, the various groups in Kabul often made little effort to coordinate their activities. These disagreements allowed Afghan communists, themselves divided, to play the sides off against one another and even to develop a lobby for their views in Moscow. Some, like Abdul Wakil, tried to sabotage the signing of the Geneva Accords in order to delay the Soviet withdrawal; others sabotaged Soviet efforts to reach out to rebel commanders. The United States faced similar problems with its South Vietnamese clients, not least President Nguyen Van Thieu, who did his best to block a US agreement with Vietnam that would end direct US military involvement.

Finally, the Soviet-Afghan conflict was prolonged by the heightened tensions between the Soviet Union and the United States in the 1980s. Soviet leaders believed that a settlement on Afghanistan would be possible only if the United States agreed to stop supporting the *mujahadeen.* At the same time, Moscow was cautious in opening a dialogue with the United States, fearing that to do so would be an admission that the invasion was a mistake and that it would lose the freedom to act as it saw fit in Afghanistan. Ultimately, however, Moscow reached out to the United States and Pakistan in hopes of reaching an international settlement that would put an end to the conflict and allow Soviet troops to come home. Similarly, Nixon and Kissinger turned to active diplomacy with Moscow and China, the patrons of Vietnam, in the hopes that a change in the international situation would help to stabilize the situation in Indochina and allow US troops to come home.

Soviet and American interventions during the Cold War were not just about setting limits or drawing lines in the sand for the rival superpower. Leaders in Moscow and Washington undertook interventions to support elites who had declared themselves for one or another version of modernity. Elites like the Khalqists, who came to power in 1978, envisaged radical transformation in their countries. When these elites found that their vision was encountering strong resistance, they called for help. Here was the culmination of the "tragedy of Cold War history," as Odd Arne Westad puts it, for nowhere was the anti-imperialist USSR more of a colonial power than in Afghanistan.[37] That this intervention was largely "colonialism by invitation" was ultimately of little comfort to the Afghan villagers trying to survive aerial bombardment or to Soviet soldiers and their families. And the Afghan government's requests for troops were easily ignored by the USSR's enemies, who used the intervention as proof that Moscow was not, in fact, an anticolonial power, but an aggressive, militaristic, and imperialist one.

It remains to evaluate Gorbachev's handling of the Afghan problem from the time he took office. Whatever the influence of officers, advisers, or other Politburo members, his was the last and most decisive word on foreign policy, at least until late 1991. He has been attacked both for betraying "friends" and for not pulling Soviet troops out earlier. There is some justification to these criticisms. If one approaches the issue dispassionately, however, it becomes clear that Gorbachev's overall approach to the Afghan problem flowed logically from the prerogatives that were largely set by the situation he inherited in 1985.

Gorbachev believed that the invasion had been a mistake and that the war had to be stopped. If it was allowed to drag on, it would remain an obstacle to his other foreign-policy aims. If he withdrew too quickly and the regime collapsed, he would quickly face criticism from many conservatives. Gorbachev could not withdraw Soviet troops before 1988, because it seemed likely that if he had done so the Kabul regime would have collapsed. The Afghan army had been able to take a lead in operations only in 1986, demonstrating its potential to operate once Soviet troops had gone. Prior to this time, it had played only a supporting role in Soviet-led battles. Babrak Karmal proved unable or unwilling either to overcome divisions within his party or to reach out to rebel commanders. Changing the Soviet strategy and giving a new Afghan leader the chance to establish himself took time, as did trying to find a diplomatic solution. As Chapter 4 has shown, Moscow worked to secure Washington's agreement to end US arms supplies to the *mujahadeen*; and after the Washington summit, Gorbachev thought he had secured that agreement.

Some scholars believe that the Afghan war played a significant role in the ultimate collapse of the Soviet state.[38] While it is true that the war helped to expose many of the injustices of Soviet military life, as well as some of the shortcomings of the military in general, it did not critically undermine the military as an all-Soviet institution.[39] The growing chorus of criticism that the military faced in the 1989–1991 period was the result more of new openness and general disillusionment with the party and state than of the war as such. Finally, it is worth remembering that the Najibullah regime collapsed four months *after* the USSR ceased to exist. Therefore, the withdrawal of Soviet troops never became a military defeat.

Although the general outline of Gorbachev's Afghan policy is understandable, even commendable, certain aspects of the way he handled it deserve serious criticism. His most important failing is that he never really took control of the Afghan problem. He trusted his deputies and colleagues to follow the general line. Usually this approach is considered smart management, but the conflict had created too many internal conflicts that reached to the top of the Soviet leadership, and it is clear that by 1987–1988 there were widely differing interpretations of National Reconciliation, the extent of the withdrawal, and the future of Soviet-Afghan relations. These differences were evident not just in policy making, where they ostensibly contributed to healthy debate, but in policy implementation, where the result was contradictory and often conflicting endeavors. This fatally undermined attempts to provide real peace and reconciliation for a much-scarred nation, a legacy that resonates to the present day.

There are many differences between the Soviet experience in Afghanistan and the situation that has unfolded since 2001, but the similarities are nevertheless remarkable. Barack Obama, like Gorbachev in the 1980s, is a young reformist leader with a mandate for change who recognizes that the Afghanistan problem can seriously hamper what he wants to do elsewhere, whether in foreign or domestic policy. The dilemmas that his administration has faced are also familiar. Should the United States commit to a reinvigorated nation-building program, with a focus on building Afghan government institutions? Should it pull out its advisers and focus instead on building up a strong leader in Kabul—a policy that one think-tank has nicknamed "find the right Pashtun"?[40] Hamid Karzai, who took power after US forces helped to topple the Taliban in

2001, now seems cast in the role of Babrak Karmal—distrusted by his patrons and by his countrymen, isolated and with little influence even over his supporters. Should Washington continue to support him, or should it look for a new face to lead Afghanistan, someone perhaps with more legitimacy among the Afghan people?[41] If troops are withdrawn soon and the Karzai government collapses, will US and UK leaders be able to explain this to their own people? Will they be able to justify continued intervention, even as casualties mount?[42]

The reality is that the Afghanistan problem is even more of a challenge for the United States today than it was for Moscow in the 1980s. For one thing, whereas Moscow was trying to stop a functioning state from disintegrating, since 2001 the task has been to build that state almost from scratch. The difficulty of coordinating the efforts of various agencies working in Afghanistan is compounded today by the problem of coordinating NATO allies—each of which has its own limits, mandates, and domestic political pressures—and the various NGOs. On the surface, at least, the regional context should be more favorable: Pakistan is a nominal ally, the Central Asian states that border Afghanistan have provided support to the US war effort, and even Iran has proved willing to cooperate. But Pakistan is an unstable (possibly failing) state with nuclear weapons, a public that does not trust the United States, and an intelligence service which is known to be supporting the Taliban and other extremist groups; in the words of vice president Joseph Biden's national security adviser, with the ISI "there is appeasement one day, confrontation another day, and direction a third day."[43] Relations with Iran deteriorated under George W. Bush and have not recovered since; the Central Asian states are unstable autocracies whose people are ambivalent about the US presence in the region.

The one bright spot may be Russia. Although relations between Moscow and Washington have often been strained since 2001, there is little doubt that Russian officials have no desire to see the NATO/ISAF effort collapse. Despite lingering bitterness over the US role in the 1980s and unhappiness with US policies under George W. Bush, there is a recognition that a failed state in Afghanistan is arguably an even more immediate threat to Russia than it is to the United States or its European allies. In an op-ed essay published in the *New York Times* in January 2010, not long after President Obama's strategy for Afghanistan was announced,

Boris Gromov, now the governor of the Moscow region, and Dmitry Rogozin, Russia's ambassador to NATO, chided Western politicians who were willing to contemplate defeat in Afghanistan:

> A rapid slide into chaos awaits Afghanistan and its neighbors if NATO pulls out, pretending to have achieved its goals. A pull-out would give a tremendous boost to Islamic militants, destabilize the Central Asian republics, and set off flows of refugees, including many thousands to Europe and Russia. . . . The minimum that we require from NATO is consolidating a stable political regime in the country and preventing Talibanization of the entire region. . . . We are ready to help NATO implement its UN Security Council mandate in Afghanistan. We are utterly dissatisfied with the mood of capitulation at NATO headquarters, be it under the cover of "humanistic pacifism" or pragmatism.[44]

These were striking words from the commander who oversaw the Soviet withdrawal and from a diplomat known for his anti-NATO and anti-US outbursts. President Obama's "reset" of relations with Moscow has been quite useful—Russia now helps to supply NATO troops in Afghanistan, and it has even taken part in anti-narcotics operations in the country.[45]

Central to the Obama administration's challenge has been trying to define victory. Is success the creation of a democratic, Western-oriented Afghanistan, as the Bush administration had envisioned? Is it the formation of a state free of the Taliban, or simply of Al-Qaeda? In fact, the definition of success has been narrowed: from "defeating" the Taliban to "degrading" the Taliban; from "democracy" to "stability." Even as the goals become less ambitious, the likely withdrawal date recedes. The deadline that the administration set in December 2009 for a draw-down of US troops—a deadline of July 2011—has already been largely abandoned. Now it is expected that US troops will have to stay in the country at their current level (about 100,000) at least through 2014. But even if—as the administration hopes—the extra time and troop levels will allow US and NATO troops to degrade the Taliban while the Afghan National Army and Afghan National Police develop into reliable security forces, a "Najibullah" scenario seems unlikely. Hamid Karzai is too weak to hold on to power, and there has so far been no evidence that a viable alternative leader exists. Meanwhile, the failure to reconcile Karzai with his electoral opponent, Dr. Abdullah Abdullah (a Tajik who was close to Ahmad Shah Massoud), bodes ill for relations among the country's ethnic groups, who are increasingly turning inward as faith in the central government collapses. It is not surprising that at least one former US official is now advocating de facto partition of Afghanistan into two, with NATO forces undertaking counterterrorism in the Pashtun-dominated south and east, and nation building in the rest of the country.[46]

Interventions become tragedies not only for the civilians caught up in conflict and the soldiers sent to fight, but also for the intervening powers themselves. Leaders who inherit the interventions, like Gorbachev and Obama—both of whom were reformers bringing a promise of change to nations in crisis—must take care not to compound previous errors out of a fear that they will be seen as weaklings unable to defend the interests of their countries. None of which is to say that peace and stability—or some measure of both—is impossible in Afghanistan. On the contrary, Afghanistan has enjoyed long periods of peace amid a number of destructive internal conflicts in its history. In the end, it will be up to the Afghan and Pakistani leaders to build that peace.

ABBREVIATIONS

APRF	Archive of the President of the Russian Federation, Moscow
AVPRF	Archive of the Foreign Ministry of the Russian Federation, Moscow
CC CPSU	Central Committee of the Communist Party of the Soviet Union
CRA	Council on Religious Affairs, USSR
CDSP	*Current Digest of the Soviet Press*
CWIHP	Cold War International History Project
CWIHP Afghanistan	Cold War International History Project, Virtual Archive: Documents on Afghanistan, 1978–1989
CWIHP Bulletin	*Cold War International History Project Bulletin*
DRA	Democratic Republic of Afghanistan
DYOA	Democratic Youth Organization of Afghanistan
GARF	State Archive of the Russian Federation, Moscow
GFA	Archive of the Gorbachev Foundation, Moscow
GFA CD	Cherniaev Diary, Gorbachev Foundation
GFA PB	Politburo meeting notes, Gorbachev Foundation
GRU	*Glavnoe Razvedyvatel'noe Upravlenie:* Main Intelligence Directorate (Soviet Military Intelligence)
KhAD	State Information Services (DRA)
IMEMO	Institute for World Economy and International Relations
NSA	National Security Archive, Washington, DC

NSA Afghanistan	Afghanistan Collection: Documents collected from US and Russian archives, National Security Archive, Washington, DC
NSA End of Cold War	"End of the Cold War" Collection: Documents collected from Russian Archives, National Security Archives, Washington, DC
NSA READ/RADD	READ/RADD Collection: Documents collected from Russian archives, National Security Archive, Washington, DC
PDPA	People's Democratic Party of Afghanistan
Pérez de Cuéllar Papers	Papers of UN Secretary General Javier Pérez de Cuéllar, Sterling Memorial Library, Yale University, New Haven, CT
RA	Republic of Afghanistan
RGAE	Russian State Archive of the Economy, Moscow
RGANI	Russian State Archive of Contemporary History, Moscow
SAM	Surface-to-Air Missile
SML	Sterling Memorial Library, Yale University, New Haven, CT
TASS	Telegraph Agency of the Soviet Union
UNA	Archives of the United Nations, New York, NY
UNGOMAP	United Nations Good Offices Mission in Afghanistan and Pakistan
WAD	Ministry of State Security (Afghanistan)

NOTES

Introduction

1. In the words of Karen Brutents, former deputy of the International Department of the CC CPSU (Central Committee of the Communist Party of the Soviet Union), there was a certain "logical progression" in the decisions to intervene or provide significant military aid first in Angola, then in Ethiopia, and then in Afghanistan. The seeming success of the first two interventions, and the general climate of Cold War confrontation in the late 1970s, both set the context for the decision to intervene in Afghanistan. Georgii Kornienko, a deputy foreign minister who also took part in the discussion, agreed with Brutents: "This competition of superpowers had its own logic. . . . Angola, it's okay. Why not Ethiopia? It's okay. Just as Czechoslovakia defined what we could do in Europe." See Westad, ed., "US-Soviet Relations and Soviet Foreign Policy toward the Middle East and Africa in the 1970s," 49–52.
2. Zubok, *A Failed Empire*, 267.
3. The closest we have is Liakhovskii, *Tragediia i doblest' Afgana.* Although the invasion of Afghanistan by NATO forces in 2001 has renewed interest in the Soviet experience there, there have been few attempts at overviews of the Soviet war effort. The exceptions largely ignore the politics, diplomacy, and decision making behind the war, and focus instead on the experience of soldiers and military operations in general. See, for example, Feifer, *The Great Gamble.*
4. Writing on the Afghan war began soon after the introduction of Soviet troops. The invasion was sharply criticized by both the left and the right in the West, and this was reflected in contemporary accounts by specialists on the Soviet Union and Afghanistan. It was often assumed that the invasion was part of an attempt to spread Soviet influence and bring it closer to the Persian Gulf, although a number of more nuanced accounts challenged this view. See Shipler, "Out of Afghanistan," for a survey of the literature from this period. Among those who noted the defensive nature of the intervention was Harry Gelman, though he also believed that Soviet leaders saw the Afghan revolution as an op-

portunity to spread their influence southward. See Gelman, *The Brezhnev Politburo and the Decline of Détente,* 170–171.

5. Westad, *The Global Cold War,* 68–72, 300. On Khrushchev's trip to Afghanistan and the resulting aid package, see Fursenko and Naftali, *Khrushchev's Cold War,* 81–82.
6. Examples are Ethiopia, where Moscow backed the Mengistu government against the Chinese-backed Tigrayan People's Liberation Front, and Angola, where China was one of the backers of UNITA. China was also a backer, though a relatively minor one, of the Afghan resistance.
7. Campbell and MacFarlane, eds., *Gorbachev's Third World Dilemmas,* 73. These numbers are based on Western calculations; it should be noted that the equipment sent in these cases was often second-hand.
8. Pikhoia, *Sovetskii Soiuz,* 331–335; Zubok, *Failed Empire,* 201–256.
9. Wilentz, *Age of Reagan,* 48–72; Del Pero, *The Eccentric Realist,* 111–144.
10. Wilentz, *Age of Reagan,* 100–125; Melvyn P. Leffler, *For the Soul of Mankind,* 259–337; Kaufman and Kaufman, *The Presidency of James Earl Carter,* 43–51; Dan Caldwell, "US Domestic Politics and the Demise of Détente," in Westad, ed. *The Fall of Détente,* 95–117; Robert Strong, *Working in the World,* 98–122.
11. Bose and Jalal, *Modern South Asia,* 220–234.
12. Dorronsoro, *Revolution Unending, Afghanistan,* 64–65; Barfield, *Afghanistan: A Cultural and Political History,* 216–225.
13. The border between Afghanistan and Pakistan, the so-called Durand Line, was indeed a colonial holdover, the result of an agreement between Amir Abdur Rahman and Sir Mortimer Durand, a British colonial official, in 1893.
14. Quoted in Kux, *The United States and Pakistan,* 245.
15. Ibid., 245–291. Talbott, *Pakistan: A Modern History,* 245–283.
16. Schulze, "The Rise of Political Islam, 1928–2000"; Esposito, ed., *Political Islam,* 1–14; Olivier Roy, *The Failure of Political Islam,* 1–27, 147–167.
17. Quoted in Westad, *Global Cold War,* 91.
18. Prior to the revolution, peasants paid virtually no tax to the state; the government's revenue came primarily from the sale of natural gas and from taxes on foreign trade; indirect taxes in the decade prior to revolution were 7–9 percent of the government's revenue. Rubin, *Fragmentation of Afghanistan,* 62–75.
19. "Decree of the CC CPSU: An Appeal to the Leaders of the PDPA Groups 'Parcham' and 'Khalq,'" January 8, 1974, *CWIHP Virtual Archive: Soviet Invasion of Afghanistan,* Cold War International History Project; Vasilii Mitrokhin, "The KGB in Afghanistan," *CWIHP Working Paper Series,* p. 26.
20. The aid Afghans accepted in this period and the projects embarked upon largely excluded major high-modernist experiments. Among the exceptions

was the Helmand Valley Project—an irrigation and settlement experiment modeled somewhat on the Tennessee Valley Authority (TVA). See Cullather, "Damming Afghanistan."

21. Author's interview with Seraya Baha, April 6, 2010.
22. Cullather, "Damming Afghanistan," 515–520. The best overview of the building of the modern Afghan state, a study that is still unsurpassed, is Gregorian, *The Emergence of Modern Afghanistan.* A most welcome recent overview is Thomas Barfield, *Afghanistan: A Cultural and Political History.* On the reforms of King Amanullah (ruled 1919–1929) and his fate when he attempted more radical modernization, see Poullada, *Reform and Rebellion in Afghanistan,* and Barfield, *Afghanistan,* 174–195.
23. Rubin, *Fragmentation of Afghanistan,* 116–117.
24. On activism prior to the revolution, see Roy, *Islam and Resistance in Afghanistan,* 69–83. See also Olesen, *Islam and Politics in Afghanistan,* 227–255; and Roy, *The Failure of Political Islam,* 147–167.
25. Author's interview with Ahmad Muslem Hayat, London, June 2010; and with Abdul Razaq, Moscow, July 2010.
26. The realist scholar Hans Morgenthau does note, however, that ideology continued to motivate US interventions during the Cold War; see his classic 1967 article, "To Intervene or Not Intervene?" See also Taliaferro, *Balancing Risks,* 3–7, for a more limited interpretation. On the limits the Cold War imposed on US interventions, see Bull, "Intervention in the Third World."
27. Westad, *Global Cold War,* 5.
28. Latham, "Nation-Building in South Vietnam," 34–35.
29. Westad, "Concerning the Situation in A."; Gibbs, "Reassessing Soviet Motives for Invading Afghanistan"; Wolf, "Stumbling towards War."

1. The Reluctant Intervention

1. Author's interview with Seraya Baha, April 6, 2010.
2. Andrew and Mitrokhin, *The World Was Going Our Way,* 388–389.
3. Ibid., 389.
4. Information from the CC CPSU to the leader of the German Democratic Republic (GDR), Erich Honecker, October 13, 1978, in CWIHP Virtual Archive: *Soviet Invasion of Afghanistan,* hereafter CWIHP *Afghanistan.*
5. Ibid.
6. Andrew and Mitrokhin, *The World Was Going Our Way,* 390–391.
7. Liakhovskii, *Tragediia i doblest',* 111.
8. Transcript of CC CPSU Politburo discussions on Afghanistan regarding deterioration of conditions in Afghanistan and possible responses from the Soviet Union, March 17, 1979, CWIHP *Afghanistan.*

9. Zubok, A *Failed Empire*, 260–261; Westad, *Global Cold War*, 288–330; Brutents, *Tridsat' let na Staroi Ploshadi*, 465. Andrei Aleksandrov-Agentov was a long-serving foreign-policy aide of Brezhnev's. He was particularly important because the General Secretary, with little knowledge of foreign affairs yet carrying enormous responsibility, relied on someone to interpret the problems and the proposed solutions brought before him. In his memoirs, Aleksandrov-Agentov himself barely mentions his own involvement with deliberations on Afghanistan. He claims to have learned of the invasion after the fact—but this claim does not exclude the possibility that he was involved in the buildup throughout 1979. See Aleksandrov-Agentov, *Ot Kollontai do Gorbacheva*, 246–247.
10. Ponomarev was a member of the commission, but did not have the same clout as Andropov, Gromyko, and Ustinov, nor the same proximity to Brezhnev. When Andropov and Ustinov began pushing for intervention, they effectively side-stepped Ponomarev.
11. From documents found by Anatolii Dobrynin and read into the record of the Lysebu II conference, 77. Hereafter Lysebu II.
12. Zubok, A *Failed Empire*, 261–262. Gai and Snegirev, *Vtorzhenie*, 204–208.
13. Cable from Soviet foreign minister Gromyko to Soviet representatives in Kabul, September 15, 1979, *CWIHP Afghanistan*.
14. Excerpt from Politburo meeting, September 20, 1979, *CWIHP Afghanistan*.
15. Gromyko-Andropov-Ustinov-Ponomarev Report to CC CPSU, October 29, 1979, *CWIHP Afghanistan*.
16. Lysebu II, 64–65.
17. This memorandum was uncovered by Anatolii Dobrynin and read into the record of the Lysebu conference. See Lysebu II, 78.
18. Ibid., 79.
19. Ibid.
20. Westad, "Concerning the Situation in A.," 29.
21. Soon after Taraki's murder, Andropov told Viacheslav Kevorkov, a senior KGB officer who served as a backchannel between Moscow and Bonn, of his concern that the United States would use Amin to pull the Soviet Union into "another Vietnam." This perhaps explains Andropov's reluctance to commit militarily in the way Ustinov wanted. Andropov did not believe that Amin could be allowed to stay in power, but he was still aware of the potential costs of an intervention. See Kevorkov, *Tainyi Kanal*, 244–245.
22. Lysebu II, 64–66.
23. Kornienko and Akhromeev, *Glazami Marshala i Diplomata*, 26.
24. Lysebu II, 74.
25. Ustinov made the connection between Afghanistan and the missiles in Germany in a conversation with Viacheslav Kevorkov. Ustinov noted that the

United States might take advantage of a vacuum in Afghanistan, creating yet another US base along the Soviet border (others being in Greece, Turkey, and Pakistan). The same was true, he went on, with regard to West Germany. If the United States placed the new missiles there, then the USSR would have to "find a way to reestablish the balance." See Kevorkov, *Tainyi Kanal*, 235.

26. Giscard d'Estaing, *Le Pouvoir et la vie*, 427.
27. According to Karen Brutents, when Aleksandrov-Agentov learned that Brutents was writing a memorandum arguing against intervention, he said, "So, do you suggest giving Afghanistan to the Americans?" Brutents' memorandum was excluded from materials presented to the Politburo. Westad, "Concerning the Situation in A.," 131.
28. Zubok, A *Failed Empire*, 262–264; Westad, "Concerning the Situation in A.," 128–132. Westad also points out that Alexei Kosygin, who had voiced his opposition to an intervention in March, was absent from the key Politburo meeting on December 12, thus removing a crucial restraining voice.
29. Artem Krechetnikov, "Afghanistan: The Soviet Vietnam," BBC Russia, June 20, 2007.
30. Liakhovskii, *Tragediia i doblest'*, 356; Dobrynin, *Sugubo Doveritel'no*, 469.
31. Such parallels are drawn in, for example, Douglas MacEachin, *Predicting the Soviet Invasion of Afghanistan.*
32. Harrison and Cordovez, *Out of Afghanistan*, 58.
33. Liakhovskii, *Tragediia i doblest'*, 355.
34. Gai and Snegirev, *Vtorzhenie*, 113.
35. Liakhovskii, *Tragediia i doblest'*, 352. Gai and Snegirev, *Vtorzhenie*, 113, confirms that the population seemed to welcome the Soviet troops at first.
36. Giustozzi, *War, Politics and Society in Afghanistan*, 10.
37. Gromov, *Ogranichennyi Kontingent*, 118.
38. Record of Politburo Meeting, January 17, 1980, in Liakhovskii, *Tragediia i doblest'*, 334. Archival reference APRF Fond 3, Opis 120, Delo 44.
39. Politburo Meeting, February 7, 1980, in Liakhovskii, *Tragediia i doblest'*, 357. APRF Fond 3, Opis 120, Delo 44.
40. Kakar, *Afghanistan*, 114–116.
41. Gromov, *Ogranichennyi Kontingent*, 118.
42. Gai and Snegirev, *Vtorzhenie*, 116.
43. Ibid.; and Liakhovskii, *Tragediia i doblest'*, 358
44. Harrison and Cordovez, *Out of Afghanistan*, 59.
45. "Nasha bol'—Afganistan" [Our Pain—Afghanistan], Interview with Yurii Gankovskii, *Aziia i Afrika Segodnia* 6 (1989), 4.
46. Afghanistan Commission of the CC CPSU Memorandum, March 10, 1988, RGANI Fond 89, Perechen 34, Delo 5; and in Allan et al., *Sowjetische Geheimdokumente*, 304.

47. Liakhovskii writes that the document, from the archive of the General Staff, is still classified and thus cannot be quoted. Liakhovskii, *Tragediia i doblest'*, 358.
48. Politburo Meeting, February 7, 1980, in Liakhovskii, *Tragediia i doblest'*, 357.
49. "The Situation around Afghanistan and the Role of Soviet Troops," in Liakhovskii, *Tragediia i doblest'*, 359. Archival reference: APRF Fond 3, Opis 120, Delo 144.
50. Afghanistan Commission of the CC CPSU Memorandum, March 10, 1988.
51. "The Situation around Afghanistan and the Role of Soviet Troops," in Liakhovskii, *Tragediia i doblest'*, 361.
52. Speech by Andrei Gromyko, Plenum of the Central Committee of the CPSU, June 23, 1980, in Allan et al., *Sowjetische Geheimdokumente*, 382.
53. Resolution of the CC CPSU "Regarding the International Position and Foreign Policy of the Soviet Union," June 23, 1980, in Allan et al., *Sowjetische Geheimdokumente*, 384.
54. According to several people who worked for Ponomarev in the International Department, he was supposedly against the initial invasion. In any case, he put his name to the proposals that came from the Afghanistan Commission. Ponomarev was not nearly as influential as Andropov, Ustinov, and Gromyko; and when he did object to their policies, he was either overruled or bullied into accepting the troika's point of view. See Cherniaev, *Sovestny Iskhod: Dnevnik Dvukh Epokh.*
55. Giustozzi, *War, Politics and Society in Afghanistan*, 22–29; author's interview with Sultan Ali Keshtmand, London, June 22, 2009; USSR Embassy, Kabul, to CC CPSU, "Regarding Economic Conditions in Afghanistan," RGANI Fond 5, Opis 77, Delo 800, 131–135, September 15, 1980.
56. "Record of Conversation between I. T. Grishin, USSR Deputy Minister of Foreign Trade, and M. H. Jalalar, DRA Minister of Trade," August 6, 1980. RGAE Fond 413, Opis 2, Delo 739.
57. "Record of Conversation between I. T. Grishin and M. H. Mangal, DRA Ambassador to the USSR," October 14, 1982. RGAE Fond 413, Opis 2, Delo 2214.
58. Memorandum of Conversation between USSR Minister of Trade I. Aristov and M. H. Jalalar, February 13, 1986. RGAE Fond 413, Opis 32, Delo 4607; Survey of Soviet trade with Asian countries, January 10, 1986. RGAE Fond 413, Opis 2, Delo 4677.
59. CC CPSU Memorandum, "Regarding Further Measures . . . in Connection with Events in Afghanistan," in RGANI Fond 89, Perechen 34, 3. Liakhovskii, *Tragediia i doblest'*, 344.
60. Liakhovskii, *Tragediia i doblest'*, 348, 350.
61. Ibid., 350.
62. Andrew and Mitrokhin, *The World Was Going Our Way*, 407.

63. The role played by advisers is discussed in greater detail in Kalinovsky, "The Blind Leading the Blind."
64. Zharov, "Sleptsi, navizivavshie sebia v povodyri," 29. A report from 1981 notes that it was common practice for Keshtmand to meet with Soviet advisers to discuss the agenda prior to his weekly meeting with the DRA Council of Ministers executive committee. The report also noted that the plans adopted by the council included those proposed by advisers. "Regarding the work of the DRA Council of Ministers," February 18, 1982. RGANI Fond 5, Opis 88, Delo 945, 24–29.
65. Elvartynov, *Afganistan glazami ochevidtsa*, 15. See also Salnikov, *Kandahar: Zapiski Sovetnika Posolstva;* Urnov interview, Moscow, March 25, 2008. Salnikov served as a political adviser in Kandahar in the mid-1980s.
66. Including Yuli Vorontsov and Nikolai Kozyrev, two of the most influential officials, but not Fikriat Tabeev, the ambassador from 1980 to 1986.
67. Plastun and Adrianov, *Nadzhibulla: Afghanistan v Tiskakh Geopolitiki*, 68. Author's interviews with Ambassador Yuli Vorontsov, Moscow, September 11, 2007, and with Leonid Shebarshin, Moscow, September 17, 2007. Shebarshin was a career officer who went on to head foreign intelligence within the KGB under Gorbachev.
68. Kirpichenko, *Razvedka*, 360.
69. Mitochkin, *Afganskie Zapiski*, 66–67. Mitochkin, a KGB officer, served as an adviser in Afghanistan.
70. Interview with Yulii Vorontsov, Moscow, September 11, 2007.
71. "Report on the Condition of the PDPA," 1983, Personal Archive of Marshal Sokolov. Provided to the author by General Aleksandr Liakhovskii.
72. Hammond, *Red Flag over Afghanistan*, 152.
73. Nablandiants, *Zapiski Vostokoveda*, 113.
74. I. Schedrov's note on the situation in Afghanistan, forwarded from *Pravda* to CC CPSU, November 1981. RGANI Fond 5, Opis 84, Delo 855, 53.
75. Elvartynov, *Afganistan glazami ochedvitsa*, 21–23.
76. Mitochkin, *Afganskie Zapiski*, 79–80.
77. Andrew and Mitrokhin, *The World Was Going Our Way*, 408.
78. According to Giustozzi, 60–70 percent of PDPA members in the army were Khalqi. Giustozzi, *War, Politics, and Society*, 82.
79. Gai and Snegirev, *Vtorzhenie*, 195.
80. Author's interview with Leonid Shebarshin, Moscow, September 17, 2007.
81. Meeting with party advisers, handwritten notes, March 31, 1984, Personal Archive of Marshal Sokolov. Provided to the author by General Aleksandr Liakhovskii.
82. For a broader overview, see Giustozzi, *War, Politics, and Society*, 36–40.

83. Liakhovskii, *Tragediia i doblest'*, 744.
84. Gromov, *Ogranichennyi Kontingent*, 330–332.
85. See Alexiev, *Inside the Soviet Army in Afghanistan*, 15.
86. Merimskii, "Afganistan: Uroki i vivody."
87. Notes of a meeting with party advisers, March 31, 1984, Private Archive of Marshal Sokolov. Provided to the author by General Aleksandr Liakhovskii. That Sokolov was apparently ready to admit to his subordinates and colleagues in Afghanistan that the border issue could not be resolved, but found it difficult to tell the defense minister the same, is indicative of some of the communication problems even near the top of the Soviet hierarchy. It seems to confirm that even as Soviet leaders learned about the difficulties of the conflict, they were sometimes left ignorant of the complete picture.
88. Not surprisingly, the vast majority of Soviet military operations took place near the border with Pakistan, in regions such as Kandahar. Some of these were efforts to disrupt the *mujahadeen* supply chain, but others were engagements with entrenched opposition groups. The distribution of fighting within Afghanistan as seen by Soviet military planners is well illustrated by the maps provided in Maiorov, *Pravda ob Afganistane*.
89. See Westermann, "The Limits of Soviet Airpower," 41–43.
90. Sarin and Dvoretskii, *The Afghan Syndrome*, 120.
91. The highway was a relic of an earlier era of Soviet-Afghan friendship. The road was built by Soviet engineers in 1950s, a result of the USSR's push into the Third World under Premier Nikita Khrushchev. Earlier Afghan kings had resisted British and Russian offers to build communications in Afghanistan, fearing that such means of access would then be used to invade the country. See Gregorian, *The Emergence of Modern Afghanistan*.
92. See Grau and Gress, *Soviet Afghan War*, 64–67.
93. Quoted in Westermann, "The Limits of Soviet Airpower," 41.
94. Feifer, *The Great Gamble*, 100–105.
95. Sarin and Dvoretskii, *The Afghan Syndrome*, 92.
96. Lester Grau, the most prolific military analyst of Soviet fighting in Afghanistan, has suggested that the high rates of disease, quite unusual for a modern army, contributed to the falling morale of the troops and the undermining of the Soviet army's prestige within Soviet society as a whole. See Grau and Jorgensen, "Beaten by the Bugs"; and Grau and Nawroz, "The Soviet Experience in Afghanistan." It is not an unreasonable argument—the number of soldiers incapacitated by disease was over 400,000 (as opposed to the official figure for the number of wounded, released in 1989: 14,000). Nevertheless, there has as yet been no study of how this affected the military's standing in the longer term.

97. This is not to say that the adjustment was easy. The Red Army had last engaged in a similar campaign in the 1920s, in its battle against the Basmachi (Muslim anti-Soviet rebels in Central Asia), but had since largely abandoned counter-insurgency training or planning. After the invasion, Soviet generals did turn to some of the military texts written during the Basmachi campaign as they tried to reorient the 40th Army to partisan warfare in mountainous conditions. Liakhovskii, *Tragediia i doblest'*, 386–387.
98. Westermann, "Limits of Soviet Airpower," 44–50.
99. Okorokov, *Sekretnye voiny*, 185; Braithwaite, *Afgantsy*, 223.
100. Braithwaite, *Afgantsy*, 223–224; Author's interview with Ahmad Muslem Hayat, London, June 2010.
101. Grau and Nawroz, "The Soviet Experience in Afghanistan."
102. Garthoff, *Détente and Confrontation*, 966–971.
103. GosPlan USSR [State Planning] Memorandum Regarding Expenses in Afghanistan, January 1988. Volkogonov Papers, Regional File, Box 26, Reel 17. The calculation is based on the official exchange rate at the time: 65 US cents to the Soviet ruble.
104. "Soviet Military Budget: $128 Billion Bombshell," *New York Times*, May 31, 1989. According to Stanislav Menshikov, a Russian economist, the military budget represented some 20–25 percent of GDP. However, it is not clear what methodology he uses to get at this number. Menshikov, "Stsenarii Razvitia VVP," 86.
105. "Repayment of developing countries' debt," October 1991. GARF Fond 10026, Opis 5, Delo 640.
106. CC CPSU Memorandum "Regarding the completion of the withdrawal of Soviet troops from the Republic of Afghanistan," February 16, 1989. Volkogonov Papers, Regional File, Box 26, Reel 17. The memorandum makes no mention of psychological trauma. There is also some controversy about the numbers—for example, a book by the Soviet General Staff cites a figure of 26,000 dead. However, it is not clear what sources were used to reach that number or why it differs from the official tally. On this point, see Grau and Gress, *The Soviet Afghan War*, 44.
107. Halliday, "Soviet Foreign Policymaking and the Afghanistan War," 691.
108. Bearden and Risen, *The Main Enemy*, 207. For more on the US decision to supply the Afghan opposition with the Stinger, see Kuperman, "The Stinger Missile"; Crile, *Charlie Wilson's War*, 403–439.
109. Harrison and Cordovez, *Out of Afghanistan*, 194–201; Kuperman, "The Stinger Missile," esp. 244–249. See also the exchange between Kuperman and Milton Bearden, "Stinging Rebukes," in *Foreign Affairs*, January–February 2002.
110. Yousaf and Adkin, *The Bear Trap*, 199.

111. Ibid., 197–198. Another attack in Kharga using Chinese-made missiles on timed ignition destroyed a storage facility containing an estimated $250 million worth of Soviet military equipment. The video footage was replayed on television for several days. See Bearden and Risen, *The Main Enemy*, 228–231.
112. Yousaf and Adkin, *The Bear Trap*, 205.
113. Ibid., 206. See also Coll, *Ghost Wars*, 161–162; and Scott, *Deciding to Intervene*, 46–47. Attacks from *mujahadeen* near border areas were felt from time to time in Soviet territory even in 1987, with Soviet civilians dead as a result. See, for example, Artem Borovik, *The Hidden War*, 42.
114. One of the limits evidently set by Soviet planners was that the war should not be extended to Pakistan, although there was at least one strike by Soviet special forces that crossed the border, apparently without authorization. See Grau and Jalali, "Forbidden Cross-Border Vendetta." Pakistan did complain on occasion about Afghan Air Force jets violating its airspace, but there does not seem to have been an organized effort on the part of the DRA and the Soviet Union to take the war into Pakistan, even though aerial bombing on the Pakistani side of the border could have stemmed the flow of arms into Afghanistan. Soviet records of Pakistani complaints are unavailable. Complaints lodged with the UN Secretary General can be found in the UN Archives, Secretary General's papers, Problem Area: Afghanistan files.
115. The Soviet decision to limit the extent of the war seems to have been recognized by the US Army in its internal assessment in 1989. One section, unfortunately mostly redacted, is called "Limited Goals, Limited Commitment." US Army, "Lessons from the War in Afghanistan," May 1989 (Army Department Declassification Release), NSA Afghanistan: Lessons from the Last War: www.gwu.edu/~nsarchiv/NSAEBB/NSAEBB57/us.html (accessed July 22, 2009).
116. Working Record of CPSU Central Committee Politburo Meeting, July 30, 1981, published in *Krasnaya Zvezda*, February 15, 2000. Translated by Gary Goldberg. CWIHP Bulletin 14/15, 245.
117. Ibid.
118. Alexievich, *Zinky Boys*, 3.
119. For the war's effects on veterans and society, see Galeotti, *Afghanistan: The Soviet Union's Last War*.
120. Kuzio, "Opposition in the USSR," 104.
121. Cherniaev Diary, NSA, April 4, 1985.
122. See, for example, Bennigsen and Broxup, *The Islamic Threat to the Soviet Union*. Others were more skeptical. Muriel Atkin, a Central-Asia specialist, noted a certain tendency among Western scholars to "demonize" the treatment of Soviet Muslims. See Atkin, "The Islamic Revolution," 94. Fred Halliday, writing several years earlier, pointed out that the idea of an Islamic challenge

to the USSR, as developed by Western scholars, arose in part from "Cold War wishful thinking about the possible challenge to the USSR of politicised Islam, a process in which academic industry and state finance have joined enthusiastically." See Halliday, "Islam and Soviet Foreign Policy," 218.

123. In fact, Andropov recalled this episode of Soviet history at a 1983 Politburo meeting: "Miracles don't happen. Sometimes we are angry at the Afghans . . . but let's remember our fight against the Basmachi. Back then almost the entire Red Army was concentrated in Central Asia, and the fight with the Basmachi went on into the 1930s." Record of Politburo Meeting, March 10, 1983, Allan et al., *Sowjetische Geheimdokumente*, 410.
124. For example, an investigation in the Chechen-Ingush ASSR in 1979 uncovered nineteen "official" and dozens of "unofficial" recording studios that were copying and distributing such tapes, which included prayers, religious instructions, and admonishments for young men not to join the Soviet army. "Regarding the *samizdat* of ideologically improper musical compositions in the Chechen-Ingush ASSR," May 11, 1979. RGANI Fond 5, Opis 76, Delo 124. See also Ro'i, *Islam in the Soviet Union*, 426.
125. Kuzio, "Opposition in the USSR," 114.
126. Cherniaev Diary, NSA, August 27, 1985. Cherniaev read about the incident in the Central Committee Secretariat protocol.
127. V. A. Kuroedov (CRA) to the CPSU Central Committee, September 16, 1983. RGANI Fond 5, Opis 89, Delo 82, p. 60; Ro'i, *Islam in the Soviet Union*, 346.
128. Ro'i, *Islam in the Soviet Union*, 717.
129. Rashid, *Jihad*, 44.
130. Bennigsen, "Afghanistan and the Muslims," 298.
131. Coll, *Ghost Wars*, 104–105.
132. Yousaf and Adkin, *The Bear Trap*, 192–195.
133. Memorandum to CRA official V. A. Nurullaev, March 1984. GARF Fond 6991, Opis 6, Delo 2761.
134. CRA memorandum, January 19, 1982. GARF Fond 6991, Opis 6, Delo 2306.
135. Information regarding a CRA conference in Tashkent, April 18, 1984. GARF Fond 6991, Opis 6, Delo 2762.
136. "USSR: Domestic Fallout from Afghan War," a CIA Intelligence Assessment, February 1988, 4–5, www.foia.cia.gov/docs/DOC_0000500659/DOC_0000500659.pdf (accessed December 15, 2010).
137. Ibid., 12.
138. Author's interviews with Hamid Ismailov, Uzbek author and journalist, London, October 22, 2009; and with Sirojiddin Tolibov, Tajik journalist, London, November 5, 2009. See also the autobiographical account of Timur Zaripov, a Tajik from the mountainous part of the republic, in Heinämaa et al., *The Soldier's Story*, 97–103.

139. Author's interview with Mehdid Kabiri, leader of the Islamic Renaissance Party of Tajikistan, London, October 17, 2009.
140. Galeotti, *Afghanistan: The Soviet Union's Last War*, 27–29.
141. This is not to say that they did not consider certain manifestations of Islam as dangerous. There were numerous resolutions on ways of combating Islam before and after the Soviet invasion. In July 1986 the question was discussed at a Politburo meeting, where Gorbachev called Islam a "dark religion" and admitted that its influence seemed to be growing. The question was not connected to the Afghanistan war, nor did Gorbachev seem particularly alarmed. See Politburo discussion, July 24, 1986, Gorbachev Foundation Archives, Notes of Politburo Discussions (1986), 149.
142. Muriel Atkin, writing in 1989, noted that Soviet officials' public attitude regarding Islam actually changed little during the early 1980s, in comparison with the pre-invasion period. Atkin, *The Subtlest Battle*, 39–40.
143. Author's interview with Leonid Shebarshin, March 19, 2008.

2. *The Turn toward Diplomacy*

1. European reaction, however, was generally less dramatic than that of the United States. See Loth, *Overcoming the Cold War*, 160–164.
2. See Duncan, *The Soviet Union and India*, 27–31.
3. "Action by the General Assembly," UN Yearbook 1980, 301.
4. This was one of the "three obstacles" to improvement of relations with the USSR, as Chinese officials framed it; the other two were the presence of Soviet troops in Afghanistan and Soviet support for Vietnam's occupation of Cambodia. See Radchenko, *Facing the Dragons*, manuscript in progress.
5. Carter, *Keeping Faith*, 483.
6. Matlock, "The End of Détente," 12.
7. Apparently, an enraged Brezhnev was against its being referred to as a doctrine at all. Anatolii Cherniaev, diary entry February 9, 1980; Cherniaev, *Sovmestnyi iskhod*, 392–393.
8. Politburo note, "Measures for the Activation . . . ," March 13, 1980. RGANI, Fond 89, Perechen 34, Delo 7, Appendix 2.
9. Ibid.
10. Bennigsen, "Soviet Islamic Strategy after Afghanistan," 57.
11. Ibid., 58.
12. Ibid., 59.
13. "Afghanistan Situation," UN Yearbook 1981, 234.
14. Politburo Protocol 191, April 5, 1980. RGANI, Fond 89, Perechen 34, Document 7.
15. Author's interview with Leonid Shebarshin, Moscow, September 17, 2007.

16. Such approaches came from the European Economic Union, French president Valéry Giscard d'Estaing, and the government of Italy.
17. Khristoforov, "Trudny put' k Zhenevskim Soglasheniam," 27–30. Obitchkina, "L'Intervention de l'Union Sovietique en Afghanistan"; Giscard d'Estaing, *Le Pouvoir et la vie*, 386–440.
18. Record of conversation between Kornienko and French ambassador A. Froman-Meris, September 14, 1981. AVPRF Fond 71, Opis 71, Perechen 113, Delo 4, pp. 31–32.
19. "The Situation in Afghanistan and Certain Questions Arising from it," April 7, 1980. In Liakhovskii, *Tragediia i doblest'*, 376. Archival reference APRF, Fond 3, Opis 82, Delo 148.
20. Kornienko, *Kholodnaia Voina*, 249–250.
21. In particular, institutes whose function was to advise the Central Committee, such as the Institute of World Economy and International Relations (IMEMO).
22. The memorandum is excerpted, almost in full, in Liakhovskii, *Tragediia i doblest'*, 337–340. A shorter excerpt is in CWIHP Bulletin 14/15, 241–242. In fact, Bogomolov later told Gai and Snegirev that he wasn't sure the memorandum had ever reached Brezhnev's eyes. Gai and Snegirev, *Vtorzhenie*, 115.
23. Cherniaev, *Sovmestnyi Iskhod*, 388–393.
24. Vadim Kirpichenko, at the time a deputy chief of the First Directorate of the KGB, notes that even in the upper echelons of the KGB there was a sense that the invasion had been a mistake. Kirpichenko, *Razvedka*, 358.
25. Arbatov, *Chelovek Sistemi*, 292; Cherniaev diary entry for June 21, 1980, in Cherniaev, *Afganskii Vopros*, 73.
26. Kirpichenko, *Razvedka*, 358–359. See also Gankovskii, "Afghanistan: From Intervention to National Reconciliation."
27. Although there was no public demonstration by dissidents—as there had been in 1968, when prominent intellectuals like the poet Evgenii Evtushenko protested the invasion of Czechoslovakia in Red Square—some, like the physicist Andrei Sakharov, expressed their discontent in letters addressed to the Central Committee (Gai and Snegirev, *Vtorzhenie*, 100). The opinions of dissidents generally had even less of an influence on policy makers at this time than did those of intellectuals like Bogomolov or Arbatov. It was all too easy to write these petitions off as the views of the "intelligentsia," who were never to be fully trusted when matters of state interest were concerned. See Andropov's comment to Boris Ponomarev following Arbatov's interview with Brezhnev, in Cherniaev, *Afganskii vopros*, 73.
28. Letter from *Pravda* correspondent I. Schedrov to the CC CPSU on the situation in Afghanistan, November 12, 1981.

29. Ibid.
30. Kapitsa interview with Odd Arne Westad.
31. Author's interview with Nikolai Kozyrev, November 15, 2008.
32. Gankovskii, "Afghanistan: From Intervention to National Reconciliation," 133.
33. Merimskii, "Afganistan: Uroki i vivody," 29.
34. This comment was noted by Vladimir Plastun, a Soviet adviser. Plastun and Adrianov, *Nadzhibulla*, 80.
35. Maiorov, *Pravda ob Afganskoi voine.*
36. Gai and Snegirov, 204–205.
37. Harrison and Cordovez, 65. Gai, "Afganistan: Kak Eto Bylo."
38. Quoted in Harrison and Cordovez, 65.
39. Zubok, *Failed Empire*, 220–221, 245.
40. Liakhovskii, *Tragediia i doblest'*, 356.
41. Giscard d'Estaing, *Le Pouvoir et la vie*, 432–433.
42. Working Record of CPSU Central Committee Politburo Meeting, July 30, 1981, published in *Krasnaya Zvezda*, February 15, 2000. Translated by Gary Goldberg. CWIHP Bulletin 14/15, 245.
43. Ibid.
44. Safronchuk, "Afganistan Pri Babrake Karmale i Nadzhibulle," 36.
45. "Regarding Our Further Line of Foreign Policy . . . and a Reply to F. Castro," March 10, 1980, RGANI Fond 89, Perechen 34, Delo 5, in Allan et al., *Sowjetische Geheimdokumente*, 304–306.
46. Confidential Note for the Secretary General, March 27, 1980, UN Archives S-0904-0089-05-1.
47. "The Situation in Afghanistan and Certain Questions Arising from it," April 7, 1980, in Liakhovskii, *Tragediia i doblest'*, 377.
48. Notes on a meeting between Mr. Agha Shahi, Minister for Foreign Affairs of Pakistan, and the UN Secretary General, September 11, 1980, UN Archives S-0904-0089-05-1.
49. Notes on a meeting between the Secretary General and the Soviet chargé d'affaires, Richard S. Ovinnikov, January 9, 1981, UN Archives S-0904-0089-6-1.
50. Harrison and Cordovez, *Out of Afghanistan*, 77.
51. "The Soviet Position as It Emerged during the Secretary General's Visit to Moscow, May 4–7, 1981," May 21, 1981, UN Archives S-1067-1-1.
52. Memorandum from Pérez de Cuéllar to Kurt Waldheim, "Mission to Pakistan and Afghanistan," August 10, 1981, UN Archives S-0904-0089-6-1.
53. Harrison and Cordovez, *Out of Afghanistan*, 77.
54. Kornienko, *Kholodnaia Voina*, 250.
55. Harrison and Cordovez, *Out of Afghanistan*, 84.
56. Ibid.
57. Safronchuk, "Afganistan pri Babrake Karmale," 41.

58. Harrison and Cordovez, *Out of Afghanistan*, 84.
59. Safronchuk, "Afganistan pri Babrake Karmale," 37. There was, of course, no actual interference on Afghanistan's part.
60. Ibid., 38.
61. Harrison and Cordovez, *Out of Afghanistan*, 112. Khan, *Untying the Afghan Knot*, 100–102.
62. Safronchuk, "Afganistan pri Babrake Karmale," 41.
63. Note on the Secretary General's Meeting with President Brezhnev, Thursday, September 9, 1982, UN Archives S-1024-87-13.
64. Liakhovskii, *Tragediia i doblest'*, 380. I have not been able to find additional confirmation regarding this meeting, but it is consistent with the general shift in the Soviet attitude at the time, as well as with Andropov's apparent desire to hasten the end of the Soviet occupation.
65. Ibid.
66. Pérez de Cuéllar, *Pilgrimage for Peace*, 188.
67. Kornienko and Akhromeev, *Glazami Marshala i Diplomata*, 47.
68. Ibid.
69. Harrison and Cordovez, *Out of Afghanistan*, 123.
70. Ibid., 124. Kornienko, *Kholodnaia Voina*, 251. Kornienko was present at both this meeting and the September 1982 meeting with Brezhnev.
71. Although the issue would be reopened at the last minute in 1988 by the Afghan foreign minister.
72. It was, however, an issue of major importance for Pakistan, which had accepted some three million Afghan refugees. Afghan representatives claimed that many of these were nomads, and that Pakistan was using the issue as propaganda against the DRA.
73. Cordovez's Note for the Record, June 1983, UN Archives S-1024-3-1.
74. Politburo Meeting, March 10, 1983, in Allan et al., *Sowjetische Geheimdokumente*, 410. RGANI Fond 89, Perechen 42, Delo 51.
75. *Cold War*, documentary series, directed by Pat Mitchell and Jeremy Isaacs (CNN, 1998), episode 20: "Soldiers of God."
76. The appellation came from conservative columnist Charles Krauthammer. Scott, *Deciding to Intervene*, 1–2, 220–221.
77. Ibid., 43–54.
78. Reagan, *An American Life*, 299.
79. Private meeting between Secretary Haig and Minister Gromyko, September 23, 1981, Reagan Archives, William P. Clark Papers, Box 3, Ronald Reagan Library.
80. Author's interview with Riaz M. Khan, Pakistani deputy foreign minister and negotiator at Geneva, Washington, DC, August 2009.
81. Ibid.; Haig, *Caveat: Realism, Reagan, and Foreign Policy*, 104–111.

82. Private meeting between Secretary Haig and Minister Gromyko, September 28, 1981, Reagan Archives, William P. Clark Papers, Box 3, Ronald Reagan Library.
83. Ibid.
84. Haig-Gromyko conversation, January 26, 1982, William P. Clark Papers, Box 3, Ronald Reagan Library; "Record of Conversation between Soviet Ambassador F. A. Tabeev and Babrak Karmal," RGANI Fond 5, Opis 88, Delo 947, pp. 108–109.
85. Scott, *Deciding to Intervene*, 47–53.
86. On various aspects of the US propaganda campaign, see Cull, *The Cold War and the United States Information Agency*, 408, 417–418, 428, 448.
87. Grigory Romanov, quoted in Leffler, *For the Soul of Mankind*, 357–358.
88. Leffler, *For the Soul of Mankind*, 358.
89. Garthoff, *The Great Transition*, 117.
90. Harrison and Cordovez, *Out of Afghanistan*, 153.
91. Kornienko, *Kholodnaia Voina*, 251.
92. In 1984 Sokolov said that closing off the borders was something the 40th Army and DRA forces could not do, although they could try to cut off the most important routes. Meeting with party advisers, handwritten notes, March 31, 1984, Personal Archive of Marshal Sokolov.
93. Harrison and Cordovez, *Out of Afghanistan*, 177. See also Pérez de Cuéllar, *Pilgrimage for Peace*, 190–192.

3. *Gorbachev Confronts Afghanistan*

1. Cherniaev, *Six Years with Gorbachev*, 44–45.
2. See English, *Russia and the Idea of the West*, 49–80.
3. English, *Russia and the Idea of the West*, 71.
4. It will be remembered that when Oleg Bogomolov criticized the Soviet invasion of Afghanistan in January 1980, he did it from the platform of the IEMSS.
5. Brown, *The Gorbachev Factor*, 29.
6. English, *Russian and the Idea of the* West, 183; Anatolii Cherniaev, *Shest' let s Gorbachevym*, 9.
7. Brown, *Gorbachev Factor*, 221–225.
8. Mendelson, *Changing Course*, 82. Gorbachev made this comment to a group of scholars in January 1989; see *Pravda*, January 7, 1989.
9. In the last few years of Brezhnev's rule, conservative Politburo members attacked institutes like IEMSS. While Andropov, who had been the patron of New Thinkers like Arbatov in the 1960s, was also interested in reformist views, the "old guard" returned to dominance during the Chernenko interregnum. Although Brezhnev and Andropov were willing to hear the reformers' views,

and even agreed with them to some extent, the climate of the Cold War and the dominance of Old Thinking meant that the reformers' influence was very limited. Thus, for example, a major policy memorandum like the one submitted by Bogomolov in January 1980 went completely unanswered and unacknowledged.

10. Diary of Anatolii Cherniaev, April 4, 1985, posted on the National Security Archive website, www.gwu.edu/~nsarchiv/, cited hereafter as Cherniaev Diary, NSA.
11. Memorandum, "Regarding Certain Timely Measures," undated, but not earlier than 1987, GFA, 17923.
12. Wallander, "Soviet Policy toward the Third World in the 1990s," 54–55.
13. Garthoff, *Great Transition*, 270–271.
14. Margot Light, "The Evolution of Soviet Policy in the Third World" in Light, ed., *Troubled Friendships*, 17–22.
15. Savranskaya, "Gorbachev and the Third World."
16. Savranskaya, "Gorbachev and the Third World"; Mastny, "The Soviet Union's Partnership with India," 78.
17. Chebrikov never wrote a memoir and gave few interviews. In one, however, he does say that as KGB chief he tried to "follow Andropov's line." Unlike Andropov, Chebrikov had little foreign-policy experience, having spent most of his KGB career dealing with organizational and domestic issues. See his interview with Aleksandr Hinshtein, *Moskovskiy Komsomolets*, December 23, 1998.
18. Cherniaev Diary, entry for August 12, 1984. Cherniaev was present at the Politburo meeting (Cherniaev, *Sovmestny Iskhod*, 570–571). Nevertheless, that summer the Politburo did make some important decisions on US-Soviet relations —namely, to restart dialogue with the United States and to seek a meeting of heads of state (Dobrynin, *Sugubo Doveritel'no*, 513).
19. Record of meeting between M. S. Gorbachev and Chairman of the Revolutionary Council of the DRA, B. Karmal, March 14, 1985. Volkogonov Papers, Library of Congress, Regional File, Box 26, Reel 17.
20. Ibid. Karmal's comments were on the whole quite superficial, and mostly defended the progress already made by the party. In concluding, Gorbachev expressed the hope that by their next meeting the party would have some new "successes and progress" that they could discuss.
21. Gorbachev interview on the radio station Ekho Moskvy, February 15, 2009.
22. Ibid., in Liakhovskii, *Tragediia i doblest'*, 521. Interview with Marshal Sergei Akhromeev, 2RR, 1/4/12, 3. See also Cherniaev Diary, April 20, 1985. Cherniaev records that Georgii Kornienko, the deputy foreign minister, made this comment to Karen Brutents, an International Department official and personal friend.

23. Cherniaev Diary, April 7, 1985, *Sovmestny Iskhod.*
24. From M. S. Gorbachev's conversation with B. Craxi, May 29, 1985. GFA Fond 3, Opis 1, Document 4771.
25. Ibid.
26. This was Marshal Sokolov's post until he was promoted to minister of defense.
27. Report by General Valentin Varennikov, June 6, 1985, in Liakhovskii, *Tragediia i doblest'*, 513–514.
28. Ibid.
29. Cherniaev Diary, NSA, October 16, 1985. Cherniaev was not present at the conversation but saw the transcript soon afterward. See also Padishev, "Nadzhibulla, President Afganistana."
30. In fact, Karmal had already been in Moscow in July, ostensibly for medical treatment. Although he probably met with Gorbachev at least briefly, it does not seem that a substantive discussion took place. See *BBC Summary of World Broadcasts,* July 31, 1985. The meeting had probably been arranged in order for the two leaders to become better acquainted. By October, however, Gorbachev was more secure in what he wanted to say on the topic of Afghanistan, and this was reflected in his long statement to the Politburo the day after the meeting with Karmal.
31. Cherniaev Diary, NSA, October 17, 1985.
32. Padishev, "Nadzhibulla, President Afganistana."
33. Cherniaev Diary, NSA, October 17, 1985. Cherniaev was present at this Politburo meeting. See also Cherniaev, *My Six Years with Gorbachev*, 42–43.
34. Cherniaev Diary, NSA, October 17, 1985. See also Cherniaev, *My Six Years with Gorbachev*, 42–43.
35. Dobrynin was still ambassador to Washington, but was in Moscow at the time and attended the meeting. Dobrynin, *In Confidence*, 447.
36. Cherniaev, *My Six Years with Gorbachev*, 42.
37. "Why the Undeclared War against Afghanistan Is Being Waged," *Pravda*, February 14, 1985, *Current Digest of the Soviet Press*, 37, no. 7, 4.
38. See, for example, "This Kind Doesn't Surrender," *Sovetskaia Belorussia*, January 6, 1985, *Current Digest of the Soviet Press,* 37, no. 7, 5. This periodical is hereafter referred to as *CDSP.*
39. "Regarding Publication in the Mass Media of Material Relating to the Activity of the Limited Contingent of Soviet Troops in Afghanistan," CC CPSU Document and draft by Varennikov and Kirpichenko, RGANI Fond 89, Perechen 11, Document 103; Allan et al., *Geheimdokumente Sowjetische*, 414–422.
40. "Soviet TV Gives Its Viewers Rare Glimpse of Afghan War," *New York Times,* July 24, 1985.
41. Aleksandr Prokhanov, "Notes from an Armored Personnel Carrier," *Literaturnaia Gazeta*, August 28, 1985, in *CDSP* 37, no. 43, 7–9.

42. Ibid., 7.
43. *Komsomolskaia Pravda,* January 8, 1986. *CDSP* 37, no. 1, 1. See also "We Wear Masks," *Komsomolskaia Pravda,* September 24, 1985, *CDSP* 37, no. 40, 24.
44. Grau and Jalali, "The Campaign for the Caves"; Prados, *Safe for Democracy,* 486–487.
45. Okorokov, *Sekretnye voiny Sovetskogo Soiuza,* 180.
46. Shevardnadze, *The Future Belongs to Freedom,* 47. Author's interview with Eduard Shevardnadze, Tbilisi, Georgia, May 9, 2008.
47. Brown, *Gorbachev Factor,* 220–221.
48. Ibid., 92–93.
49. Zubok, *Failed Empire,* 284–286; Brown, *Gorbachev Factor,* 220–222; Garthoff, *Great Transition,* 256–260.
50. Author's interview with Soviet ambassador Yulii Vorontsov, Moscow, September 11, 2007.
51. Liakhovskii, *Tragediia i doblest',* 523. Unfortunately, the record of this meeting is unavailable.
52. Politburo meeting, February 23, 1987, GFA PB 1987, 114.
53. Cherniaev, *My Six Years with Gorbachev,* 43.
54. Mastny, "The Soviet Non-Invasion of Poland in 1980–1981," 29.
55. Zubok, *Failed Empire,* 267; notably, this time the objections of senior military officers who opposed intervention seem to have carried more weight. Ouimet, *Rise and Fall of the Brezhnev Doctrine,* 200–204.
56. Quoted in Patman, "Reagan, Gorbachev and the Emergence of 'New Political Thinking,'" 588.

4. The National Reconciliation Campaign

1. Author's interview with Andrei Grachev, International Department official, Gorbachev aide, and Kremlin press secretary from September to December 1991; London, February 1, 2008.
2. Author's interview with the Soviet ambassador to Afghanistan, Yulii Vorontsov, Moscow, September 11, 2007.
3. "GRU Dossier on B. Karmal," 1979, in Liakhovskii, *Tragediia i doblest',* 308–309.
4. Author's interview with ambassador Yulii Vorontsov, Moscow, September 11, 2007. See also "Soviet Cites Dissent on Afghan War," *Washington Post,* March 20, 1989; Varennikov's interview with journalist Artem Borovik, in *Ogonek,* March 1989, 6–7. Plastun and Adrianov disagree with the assessment of Karmal given by Varennikov, Gromov, Liakhovskii, and others, stating that the attitude of Soviet advisers and the time that Karmal had spent in opposition limited his effectiveness as a leader. See Plastun and Adrianov, *Nadzhibulla,* 70.
5. Gianni Picco (aide to and special envoy of Pérez de Cuéllar) to Javier Pérez de

Cuéllar, Pérez de Cuéllar Papers, Sterling Memorial Library, Yale University, Box 9, Folder 96. UN officials may have been reacting to some unofficial expressions of discontent heard from Soviet representatives. Author's interview with Selig Harrison, January 3, 2008.

6. Author's interviews with Leonid Shebarshin, September 17, 2007; and with Andrei Grachev, January 31, 2008. Liakhovskii, *Tragedia i doblest'*, 495.
7. Cherniaev Diary, October 17, 1985, NSA.
8. Habibia College was set up under King Habibia in 1903 to educate the children of elites according to a Western curriculum.
9. Plastun and Adrianov, *Nadzhibulla*, 127–128.
10. Dixit, *Afghan Diary*, 97.
11. Ibid., 60.
12. "Insinuations Regarding Changes in DRA Leadership," GRU report, in Liakhovskii, *Tragediia i doblest'*, 532–534. This assessment would be proven correct later, when Najibullah refused to deal with the Tajik leader Ahmad Shah Massoud, whom the Soviet military believed was the most prominent commander they could strike a deal with; Najib preferred instead to reach out to Hekmatyar, a Pashtun seen as one of the more "extremist" leaders.
13. Author's interviews with Leonid Shebarshin, September 17, 2007.
14. Plastun and Adrianov, *Nadzhibulla*, 127–128.
15. Author's interviews with Leonid Shebarshin, September 17, 2007.
16. Dobrynin, *In Confidence*, 442–443; Andrew and Mitrokhin, *The World Was Going Our* Way, 416; Shebarshin, *Ruka Moskvi*, 229–231.
17. Shebarshin, *Ruka Moskvi*, 229–231.
18. Ibid., 232–236.
19. Ibid., 236.
20. *Pravda*, May 5, 1986, in *Current Digest of the Soviet Press*, 37, no. 18, 22. This periodical is hereafter referred to as *CDSP*.
21. Plastun and Adrianov, *Nadzhibulla*, 80.
22. Record of Politburo meeting, September 25, 1986, GFA PB 1986, 171.
23. CC CPSU Memorandum, November 13, 1986, in CWIHP Virtual Archive: *Soviet Invasion of Afghanistan*, hereafter *CWIHP Afghanistan*.
24. Plastun and Adrianov, *Nadzhibulla*, 73.
25. Record of Politburo meeting, November 13, 1986, in Allan et al., *Sowjetische Geheimdokumente*, 434–450.
26. Author's interview with General Aleksandr Liakhovskii, Moscow, July 2006.
27. Ibid.
28. Notes of Politburo meeting, January 21, 1987, GFA PB 987, 60.
29. Kirpichenko, *Razvedka*, 362.
30. This came up repeatedly in my conversations with former Soviet officials. The parallel they had in mind was generally Amir Abdur Rahman Khan (ruled

1880–1901). Although he had to accept British control of Afghanistan's foreign affairs, he is remembered as a strong leader who did much to centralize authority, subdue rebellious tribes, and limit the power of traditional chiefs.

31. The term "National Reconciliation" was borrowed from the process then being implemented in post-Franco Spain; apparently, the christening took place on a flight from Kabul to Moscow. Author's interview with Leonid Shebarshin, September 17, 2007.
32. Aleksandr-Agentov, quoted in Gai and Snegirev, *Vtorzhenie*, 367.
33. "Regarding the Talks with Comrade Najib," Politburo Protocol, December 25, 1986, Volkogonov Papers, Box 26, Reel 17.
34. For example, Bulgaria, which had already provided many millions in aid, agreed in 1987 "to respond to our Soviet comrades' proposal, and respond to the PDPA's appeal to provide assistance to the PDPA's policy of national reconciliation in Afghanistan." Memorandum of the CC BCP Department of Foreign Policy and International Relations, CWIHP Documents on Afghanistan. Moscow made similar appeals to other Eastern European allies, all in the name of "furthering the goal of National Reconciliation." Presidium of the Central Committee of the Communist Party of Czechoslovakia, 35th session, May 6, 1987, CWIHP.
35. Politburo meeting, February 28, 1987, GFA PB 1987, 125.
36. See Rubin, *Fragmentation of Afghanistan*, 150, on the importance of Gorbachev's support for Najib's battle for power within the PDPA.
37. Politburo meeting, February 28, 1987, GFA PB 1987, 124; notes from telephone conversation with Comrade Najib, March 3, 1987, GFA Document 577.
38. "Report on the Condition of the PDPA," 1983, personal archive of Marshal Sokolov. Provided to the author by General Aleksandr Liakhovskii.
39. Padishev, "Nadzhibulla, President Afganistana," 23.
40. Author's interview with ambassador Yulii Vorontsov, Moscow, September 11, 2007.
41. Journalist Andrei Morozov's interview with Tabeev, December 2000, www.dailytalking.ru/interview/tabeev-fikryat-ahmedzhanovich/84/ (accessed December 16, 2010).
42. Politburo session, May 29, 1986, GFA PB 1986, 75. The Indian ambassador in Kabul, J. N. Dixit, had come to the same conclusion years earlier; describing his first meeting with Tabeev in 1982, Dixit noted in his diary that "pro-consul, potentate, gold finger could be a more apt description" than "ambassador." Dixit, *Afghan Diary*, 47.
43. Prior to 1986, Vorontsov had enjoyed a long career at the top of the Soviet diplomatic hierarchy. After earning his degree at the elite MGIMO (Moscow State Institute of International Relations), he served for many years in the Soviet mission to the United Nations, then as ambassador to India. In Afghani-

stan, he was charged not only with overseeing the implementation of National Reconciliation, but also with changing the way Soviet advisers operated.

44. Records of Politburo discussions, May 25 and June 11, 1986, GFA PB 1986.
45. Author's interview with ambassador Yulii Vorontsov, Moscow, September 11, 2007.
46. Plastun and Adrianov, *Nadzhibulla*, 83–84.
47. Braithwaite, *Afgantsy*, 208–209.
48. Plastun and Adrianov, *Nadzhibulla*, 83.
49. Meeting of political advisers, March 9, 1988, in Plastun and Adrianov, *Nadzhibulla*, 211.
50. For an excellent study of Massoud's emergence as a major resistance leader, see DeNeufville, *Ahmad Shah Massoud.*
51. Liakhovskii and Nekrasov, *Citizen, Politician, Warrior*, 153.
52. Ibid., 156–157. Massoud competed for resources and influence within the *mujahadeen* leadership, and was thus eager to have such discussions conducted clandestinely.
53. Author's interview with ambassador Yulii Vorontsov, Moscow, September 11, 2007.
54. Record of conversation with General Major Ziarmal, January 4, 1988, in Plastun and Adrianov, *Nadzhibulla*, 226–227.
55. Giustozzi, *War, Politics and Society*, 154–185; Rubin, *Fragmentation of Afghanistan*, 153–158; Dorronsoro, *Revolution Unending*, 195–197.
56. Plastun and Adrianov, *Nadzhibulla*, 78–79; "Record of Conversation with Colonel Mohammed Sarwari," December 18, 1987, ibid., 203–204. Mohammed Sarwari (not to be confused with Assadullah Sarwari) was a Parcham member and former Afghan military attaché in India.
57. "Record of Conversation with Major General A. Wakhed," December 1, 1987, ibid., 201.
58. Rubin, *Fragmentation of Afghanistan*, 150.
59. Plastun and Adrianov, *Nadzhibulla*, 76.
60. Records of Politburo discussions, June 1, 1987, GFA PB 1986, 342.
61. Cherniaev, Memorandum to Gorbachev, June 18, 1986, GFA Document 369.
62. Quoted in Patnam, "Reagan, Gorbachev and the Emergence of 'New Political Thinking,'" 588.
63. Gianni Picco to Pérez de Cuéllar, "Update on the Negotiations in Afghanistan," June 14, 1985. UN Secretary General's Files, S-1024-3-1.
64. Rubin, *The Search for Peace in Afghanistan*, 70–71.
65. Harrison and Cordovez, *Out of Afghanistan*, 214.
66. Ibid., 216. Rubin, *The Search for Peace in Afghanistan*, 70–71.
67. Cordovez to Pérez de Cuéllar, September 27, 1985; Record of meeting be-

tween Dost and Pérez de Cuéllar, September 28, 1985; Cordovez to Pérez de Cuéllar, October 31, 1985. UN Secretary General's Files, S-1024-3-1.

68. Record of conversation between Pérez de Cuéllar and Dost, November 1, 1985. UN Secretary General's Files, S-1024-3-1.
69. Politburo Notes, June 28, 1986, GFA PB 1986.
70. Ibid., July 11, 1986, GFA PB 1986, 129.
71. Defense minister Sergei Sokolov supported the decision, but pointed out: "We have experience: in 1980 we withdrew three divisions, but did not play it politically." Politburo Notes, July 11, 1986, GFA PB 1986, 129.
72. "Speech on the Occasion of Vladivostok's Being Given the Order of Lenin," in Mikhail Gorbachev, *Izbrannye rechi*, vol. 4, 10–11; *Pravda*, July 29, 1986, 1–3, in *CDSP* 38, no. 30, 8.
73. Khan, *Untying the Afghan Knot*, 145.
74. Politburo Notes, September 25, 1986, GFA PB 1986, 171. From 1980 to 1984, Soviet-Pakistani discussions mostly took place before UN General Assembly sessions, but even these were discontinued in 1984. Khan, *Untying the Afghan Knot*, 180.
75. Khan, *Untying the Afghan Knot*, 180–181.
76. Pakistani interlocutors had consulted the Alliance of Seven leaders in Peshawar prior to the meeting, and would do so again before Pakistani officials traveled to Moscow at the end of February.
77. Khan, *Untying the Afghan Knot*, 191–194.
78. Ibid., 197–199.
79. Abbas, *Pakistan's Drift into Extremism*, 121–123.
80. Author's interview with Yulii Vorontsov, Moscow, September 11, 2007.
81. Abbas, *Pakistan's Drift into Extremism*, 135–145; Rashid, *Descent into Chaos*, 39–40. Author's interview with Riaz Khan, Washington, D.C., August 3, 2009.
82. Author's interview with Riaz Khan, Washington, D.C., August 3, 2009.
83. Khan, *Untying the Afghan Knot*, 214.
84. Ibid., 237–238.
85. Politburo meeting, January 21, 1987, GFA PB 1987, 60.
86. Ibid.
87. Politburo meeting, February 23, 1987, GFA PB 1987, 114.
88. Ibid., June 11, 1987, GFA PB 1987, 342.
89. Khan, *Untying the Afghan Knot*, 184.
90. Ibid.
91. Gankovskii, "Afghanistan: From Intervention to National Reconciliation," 134.
92. Gankovskii to Cherniaev, "Regarding Measures to Settle the Conflict in Afghanistan," GFA Document 729.
93. Politburo meeting, May 21, 1987, GFA PB 1987, 309.

94. Ibid., May 22, 1987, GFA PB 1987, 319.
95. Record of conversation between Gorbachev and Najibullah, July 20, 1987, NSA READ/RADD, Box 9.
96. Record of conversation between Gorbachev and Najibullah, November 3, 1987, NSA READ/RADD, Box 9.
97. Kornienko, *Kholodnaia Voina,* 254.
98. Letter to minister of defense Dmitrii Iazov, August 13, 1987, NSA Afghanistan. Tsagolov was frustrated that his views were not being taken seriously and decided to air them openly in an interview in 1988, for which he was fired from the military.
99. Plastun and Adrianov, *Nadzhibulla,* 75.
100. Record of conversation between Gorbachev and Najibullah, July 20, 1987, NSA READ/RADD, Box 9.
101. Notes from Politburo meeting, July 23, 1987, GFA PB 1987, 429. To Gorbachev, "Karmalism" meant "elements of Marxism combined with dependence on the USSR."
102. Thoughts from Gorbachev's summer vacation, Politburo notes, GFA PB 1987, 471.
103. Notes from Politburo meeting, July 23, 1987, GFA PB 1987, 429.
104. Zubok, *Failed Empire,* 301.
105. Ibid.

5. *Engaging with the Americans*

1. Andropov identified alcohol consumption as one of the key problems for Soviet economic productivity. But the anti-alcohol campaign launched by Gorbachev had unfortunate consequences, including a rise in moonshine production and cases of alcohol poisoning. It also deprived the state budget of a key revenue source.
2. On domestic political and economic reform in particular, Gorbachev told Cherniaev he was prepared to go "far, very far." Zubok, *Failed Empire,* 301.
3. Reagan, *An American Life,* 588.
4. Wilentz, *Age of Reagan,* 166–167; see also Fischer, *The Reagan Reversal,* 2–5.
5. Leffler, *For the Soul of Mankind,* 358–365.
6. Wilentz, *Age of Reagan,* 249–253.
7. Reagan to Gorbachev, draft of private letter, December 1985, NSA; online at www.gwu.edu/~nsarchiv/NSAEBB/NSAEBB172/Doc29.pdf (accessed July 22, 2009).
8. Reagan to Gorbachev, December 16, 1985, Reagan Library, Executive Secretariat, Head-of-State File, Box 40.
9. Author's telephone interview with US ambassador Jack Matlock, January 1, 2008.

10. Cherniaev, *My Six Years with Gorbachev*, 43.
11. Harrison and Cordovez, *Out of Afghanistan*, 219.
12. Ibid., 220–241.
13. Minutes of Politburo meeting, November 13, 1986, in Allan et al., *Sowjetische Geheimdokumente*, 440.
14. *Mikhail Gorbachev's Replies to Questions Put by the Indonesian Newspaper "Merdeka,"* July 21, 1987, RIA Novosti (Moscow, 1987). As late as November 1986, Soviet officials publicly said that a Soviet withdrawal would only begin two years after it was clear interference had stopped. Gankovskii, "Afghanistan," 135.
15. Coll, *Ghost Wars*, 168.
16. Brown, *Gorbachev Factor*, 235.
17. Shultz, *Turmoil and Triumph*, 987.
18. Ibid., 1007.
19. Gates, *From the Shadows*, 425.
20. Khan, *Untying the Afghan Knot*, 233.
21. Record of conversation between M. S. Gorbachev and Najibullah, November 3, 1987, NSA READD/RADD, Box 9.
22. Shultz, *Turmoil and Triumph*, 1087; Coll, *Ghost Wars*, 177.
23. Excerpt from conversation between M. S. Gorbachev and President Reagan on Afghanistan, December 9, 1987, NSA READD/RADD, Box 9.
24. Ibid.
25. Excerpt from conversation between M. S. Gorbachev and President Reagan on Afghanistan, December 10, 1987, NSA READD/RADD, Box 9.
26. Ibid.
27. Memorandum from S. F. Akhromeev, "Afghanistan: The Position of the USSR," December 3, 1987, GFA 944.
28. Memorandum of Conversation in the Oval Office, December 9, 1987, NSA End of Cold War, Box 3. Reagan, *Reagan Diaries*, 556.
29. US Memorandum of Conversation of Working Luncheon, December 10, 1987, NSA End of Cold War, Box 3. See also the Russian record: excerpt from conversation between M. S. Gorbachev and President Reagan on Afghanistan, December 10, 1987, NSA READD/RADD, Box 9.
30. Coll, *Ghost* Wars, 177.
31. Soon after the summit, the *Washington Post* reported that undersecretary of state Michael Armacost, who was involved in the high-level talks on Afghanistan, and several other senior officials confirmed that the United States would end aid to the Afghan opposition once Soviet troops had withdrawn. "Aid to Rebels Would End with Soviet Pullout," *Washington Post*, December 14, 1987.
32. This is the opinion of the long-time Soviet ambassador to the United States,

Anatolii Dobrynin, who at the time was serving on the Afghanistan Commission of the Politburo. See Halliday, "Soviet Foreign Policymaking," 687.

33. Only a general statement on cooperation in Third World conflicts was made. See "Joint US-Soviet Summit Statement," *USSR-US Summit*, 67.
34. Braithwaite, *Afgantsy*, 213–215; Gromov, *Ogranichennyi Kontingent*, 119–122.
35. "Shevardnadze in Kabul for Talks with Najibullah, Sees Good Prospects for Soviet Troop Withdrawal within Twelve Months," *Pravda*, January 7, 1988, in *CDSP* 40,no. 1 (1988), 13.
36. "Interview with the Bakhtar News Agency," *Pravda*, January 7, 1988, in *CDSP* 40, no.1 (1988), 14.
37. Author's interview with Aleksandr Liakhovskii, Moscow, July 6, 2006.
38. Shevardnadze, *The Future Belongs to Freedom*, 69.
39. Diary notes of Politburo meeting. Vorotnikov, *A Bylo Eto Tak*, 219.
40. Shultz, *Turmoil and Triumph*, 1087.
41. Egorychev, "Afganistan stoil nam 15 milliardov dollarov v god."
42. "Notes of Meeting with Shevardnadze," December 15, 1987, Pérez de Cuéllar Papers, Box 9.
43. According to Nikolai Kozyrev, a deputy foreign minister who was the chief negotiator at Geneva, Shevardnadze felt vulnerable within the Soviet leadership. This feeling moved him to take a more conservative line on Afghanistan and also to ally himself more closely with Kriuchkov. Author's interview with Nikolai Kozyrev, Moscow, November 15, 2008.
44. Varennikov, "Sud'ba i Sovest'," 51.
45. Author's interview with Aleksandr Liakhovskii, Moscow, July 6, 2006.
46. The talks had stalled in 1988 in part because of disagreements regarding the Afghan-Pakistani border, the "Durand Line" that separated Pashtun clans into two political entities. Cordovez writes that, although the superpowers seemed to be closer than ever to an agreement at this time, Pakistan and Afghanistan were becoming more intransigent. This was not Moscow's main concern, however. After all, a commitment from the United States would really have been the key to getting Afghanistan to sign and would have isolated Pakistan as the sole obstructionist player. Harrison and Cordovez, *Out of Afghanistan*, 323. Kornienko believes that Pakistan's new stubbornness had to do with the fact that Shevardnadze encouraged Afghanistan to use Indian attitudes toward Pakistan as justification for Afghan positions at negotiations. The effect, as he puts it, was to "wave a red flag before the bull." Kornienko, "The Afghan Endeavor," 14. But Riaz M. Khan, the chief Pakistani negotiator at Geneva, believes that it had more to do with Zia ul-Haq's fear of isolation if a US-Soviet rapprochement on Afghanistan were to make Pakistan an unnecessary ally. See Khan, *Untying the Afghan Knot*, 236–237.

47. Kornienko, *Kholodnaia Voina*, 257.
48. Ibid.
49. "Statement on Afghanistan by M. S. Gorbachev," *Pravda*, February 9, 1988, in *CDSP* 40, no. 6, 1–2.
50. Harrison and Cordovez, *Out of Afghanistan*, 335.
51. Ibid., 334.
52. N. I. Kozyrev, *Zhenevskie soglashenie 1988 goda*, 21.
53. Ibid., 22.
54. Ibid.
55. "Statement on Afghanistan by M. S. Gorbachev," *Pravda*, February 9, 1988, *CDSP* 40, no. 6.
56. Record of a Politburo meeting chaired by Iakovlev, February 22, 1988, NSA READD/RADD, Box 9.
57. Harrison and Cordovez, *Out of Afghanistan*, 338.
58. Matlock, *Reagan and Gorbachev*, 287; Shultz, *Turmoil and Triumph*, 1089.
59. "Shultz Sure of Soviets' Afghanistan Pullout," *Washington Post*, February 24, 1988.
60. Author's telephone interview with Ambassador Jack Matlock, January 1, 2008.
61. Matlock, *Reagan and Gorbachev*, 286.
62. Harrison and Cordovez, *Out of Afghanistan*, 340.
63. Matlock, *Reagan and Gorbachev*, 285–286.
64. "The Afghanistan Announcement," *Washington Post*, February 9, 1988.
65. Record of conversation between M. S. Gorbachev and US secretary of state George Shultz, February 22, 1988, NSA READD/RADD, Box 9.
66. Record of conversation with K. Pant, February 11, 1988, NSA READD/RADD, Box 9.
67. Politburo meeting, March 3, 1988, GB PB 1988, 89.
68. Oberdorfer, *The Turn*, 279.
69. Cherniaev Diary, April 1, 1988, GFA CD 1988. See also the discussion between ambassador Aleksandr Belonogov and Pérez de Cuéllar on March 29, 1988. The Soviet position was that "symmetry" should be understood as the trade-off in the accords between the Soviet withdrawal and the cessation of arm supplies. Vorontsov had told the Pakistani ambassador that "the Soviet Union had no intention of entering into negotiations with the United States on this issue." As we will see, in the end the Politburo voted to sign the accords with the United States as a signatory—primarily to give the withdrawal more of an international legal framework. "Notes of a Meeting of the Secretary General and the Permanent Representative of the USSR," March 29, 1988, UNA, S-1024-2-3.
70. Egorychev, "Afganistan stoil nam 15 milliardov dollarov v god."
71. Author's interview with Aleksandr Liakhovskii, Moscow, July 3, 2006.

72. Ekedahl and Goodman, *Wars of Eduard Shevardnadze*, 185.
73. Record of a Politburo meeting chaired by Comrade A. N. Iakovlev, February 22, 1988, NSA READD/RADD, Box 9.
74. Politburo meeting March 3, 1988 GFA PB 1988, 89.
75. Ibid.
76. Politburo meeting April 1, 1988 GFA PB, 1988.
77. Cherniaev Diary, April 1, 1988, GFA CD 1988.
78. Padishev, "Nadzhibulla, President Afganistana," 20.
79. Details regarding the condition of the Afghan armed forces are beyond the scope of this study, but a report made on March 9 by General Varennikov is quite telling and worth citing. Varennikov noted that desertion was on the rise and that very little had actually been accomplished in terms of improving the Afghan army during the Soviet presence in Afghanistan. "Meeting of Political Workers and Advisers in the Officer's House of the 40th Army in Kabul," March 9, 1988, in Plastun and Adrianov, *Nadzhibulla*, 208–212.
80. Oberdorfer, *The Turn*, 281.
81. Najibullah, knowing that Gorbachev was eager to sign, asked about the possibility of leaving 10,000–15,000 troops for training Afghan soldiers and for guarding economic targets. Gorbachev left the question open but pointed out that it might be possible to do so within the framework of the Geneva agreement if the troops were sent as "advisers" who would train Afghans working with Soviet armaments: "After all, it is natural that when military technology is provided, there is a demand for help in mastering it. This is normal; everyone acts this way." While it could be argued that Gorbachev said this only to humor Najibullah and get his approval of the accords, Gorbachev's decision making in the fall of 1988 (discussed in this chapter), suggests that he really did believe in supporting Najibullah. Record of conversation of M. S. Gorbachev with the President of Afghanistan and General Secretary of the CC PDPA Najibullah, April 7, 1988, NSA READD/RADD, Box 9.
82. Record of conversation between M. S. Gorbachev and the president of Afghanistan, Najibullah, April 7, 1988, NSA READD/RADD, Box 9.
83. Wakil insisted that the main obstacle was the issue of borders—that is, the Durand Line, which all parties pledged to respect. In fact, it was probably an attempt to scuttle the accords, which Wakil feared would mean the end of his government. Author's interview with Nikolai Kozyrev, Moscow, November 14, 2008.
84. Harrison and Cordovez, *Out of Afghanistan*, 359.
85. Ibid.; author's interviews with Yulii Vorontsov, Moscow, September 11, 2007, and with Nikolai Kozyrev, Moscow, November 14, 2008. See also "Zalozhniki Istorii" (Interview with Nikolai Kozyrev), *Moskovskiy Komsomolets*, 49, March 5, 2004, 9.

86. Record of conversation between M. S. Gorbachev and the General Secretary of the Italian Communist Party, A. Natta, March 29, 1988, NSA READD/RADD, Box 9.
87. Record of telephone conversation between M. S. Gorbachev and F. Castro, April 5, 1988, GFA Document 20686; record of conversation between M. S. Gorbachev and the president of the CSSR, G. Husák (Czechoslovakia), April 12, 1988, GFA Document 20684.
88. Nina Andreeva was a Leningrad chemistry teacher who published an article entitled "I Will Not Forsake My Principles" in *Sovetskaia Rossiia.* The article was applauded by the more conservative party members. Brown, *Gorbachev Factor,* 172–175.
89. Gorbachev's meeting with the third group of Obkom secretaries, April 18, 1988, GFA PB 1988, 191.
90. Politburo meeting, April 18, 1988, Medvedev's notes, GFA PB 1988, 211.
91. Shevardnadze, *The Future Belongs to Freedom,* 69.
92. Author's interview with Aleksandr Liakhovskii, Moscow, July 6, 2006; Varennikov, "Sud'ba i Sovest," 51.
93. Politburo meeting April 1, 1988, GFA *Vestka v Politburo,* 312.
94. Gorbachev, *Memoirs,* 458.
95. Record of conversation between M. S. Gorbachev and the president of Afghanistan, Najibullah, April 7, 1988, NSA READD/RADD, Box 9.
96. Zubok, "Gorbachev and the End of the Cold War."
97. Savranskaya, "Gorbachev and the Third World."
98. On April 12, two days before the signing ceremony in Geneva, Reagan complained in his diary: "Another meeting with leaders of hard Conservative leaders [*sic*] Paul Weyrich, Gen. Graham etc. . . . As usual they had us on the wrong side in Afghanistan settlement, Mozambique, Chile & Angola. It's amazing how certain they can be when they know so d—n little of what we're really doing." Reagan, *Reagan Diaries,* 595.
99. Cherniaev Diary, April 1, 1988, GFA CD 1988. Cherniaev notes that he had attached a memorandum to the plan for supplies, urging Gorbachev to focus on making Mengistu change his approach to the Eritrean separatists. But Gorbachev simply detached the memorandum and signed the supply plan.
100. Politburo meeting, April 18, 1988, GFA PB 1988, 215.

6. *The Army Withdraws and the Politburo Debates*

1. "On Certain Measures to Reform Foreign Policy," 1986, GARF, Fond 10063, Opis 2, Delo 69.
2. Kozyrev, *Zhenevskie soglashenie,* 35.
3. Ibid., 46.
4. Politburo meeting, April 18, 1988, GFA PB, 211.

5. Ibid.
6. Liakhovskii, *Tragediia i doblest'*, 588.
7. Politburo meeting, April 18, 1988, Medvedev's notes, GFA PB 1988, 213.
8. US Embassy, Kabul, to State Department, May 21, 1988, NSA End of Cold War, Box 3.
9. Brown, *Gorbachev Factor*, 236–238.
10. Shultz, *Turmoil and Triumph*, 987.
11. Moscow Summit, Second Plenary Meeting, June 1, 1988, NSA End of Cold War, Box 3, p. 4. Reagan's comment during the summit that he no longer saw the USSR as an "evil empire" seemed to show that Gorbachev's strategy was working.
12. Ibid., 12.
13. The United Nations Good Offices Mission in Afghanistan and Pakistan, a team of fifty military observers sent to monitor the Soviet withdrawal. See Pérez de Cuéllar, *Pilgrimage for Peace*, 198–199.
14. See, for example, the MID letter passed to the UN Secretary General's Office, dated June 2, 1988, or the record of conversation between RA foreign minister Abdul Wakil and Pérez de Cuéllar, June 2, 1988; both in SML, Pérez de Cuéllar Papers, Box 10. The UN's inability (or unwillingness) to do anything about Pakistani "interference" is evident in Diego Cordovez's report on his visit to the area and investigation of Soviet-Afghan allegations. Apparently he was satisfied with Pakistan's response that Islamabad intended to follow the Geneva Accords to the letter, and that whatever violations took place were the result of three million Afghan refugees whose "legitimate political activities" Pakistan could not restrict. "Implementation of the Agreements on the Settlement of the Situation Relating to Afghanistan: Progress Report by the Representative of the Secretary-General," July 26, 1988, Pérez de Cuéllar Papers, Box 10, Folder 102; see also Diego Cordovez to S. Shah Nawaz, Pakistan's permanent representative to the UN, June 10, 1988, Pérez de Cuéllar Papers, Box 10, Folder 110. In a later letter to Abdul Wakil and Yaqub Khan, dated October 20, 1988, Cordovez insisted that the UN could not issue judgments on complaints and that the "letter and spirit" of the Geneva Accords required the parties to sort out the problem among themselves. Cordovez also begged both sides to be more selective in their complaints. Such meetings as Cordovez mentioned were envisaged in the accords; Pakistan initially rejected them, then agreed to conduct them at the chargé level. They never took place. Later reports of UNGOMAP investigations from 1988 generally either noted that Afghan/Soviet complaints regarding Pakistani activities (as well as Pakistani complaints about alleged Soviet/Afghani bombing on or near Pakistani territory) could not be investigated or did not constitute "clear violations of the Geneva Accords." UN officials admitted the difficulty of fully investigating most of these complaints, due to "in-

sufficient information and details, frequent impossibility of locating the positions mentioned, . . . difficulties of terrain, and security conditions." Note for the Secretary General, October 7, 1988, Pérez de Cuéllar Papers, Box 10, Folder 110. The fact that the UNGOMAP staff was small would likewise have made investigating a large number of complaints extremely difficult. See also "Afghanistan: Recent Developments" (Note for the SG), December 2, 1988, SML, Pérez de Cuéllar Papers, Box 10, Folder 110.

15. "Afghanistan: Continuing the Beginning," unsigned editorial, *International Affairs,* September 1988, 78–80.
16. Bearden and Risen, *The Main Enemy*, 354. See also Record of conversation between Vladimir Petrovsky, deputy foreign minister of the USSR, and Pérez de Cuéllar, October 21, 1988. Petrovsky told Pérez de Cuéllar that "there were different schools of thought in Moscow concerning developments relating to Afghanistan, and that those favoring the timely withdrawal of Soviet troops such as himself were—in the present circumstances—facing difficulties." SML, Pérez de Cuéllar Papers, Box 10, Folder 102.
17. Author's interview with ambassador Jack Matlock, January 1, 2008.
18. Michael Armacost to US Embassy, New Delhi, June 9, 1988, NSA End of Cold War, Box 3.
19. Politburo Meeting, September 18, 1988, GFA PB 1988.
20. Author's interview with ambassador Yulii Vorontsov, Moscow, September 11, 2007.
21. Rogers, *The Soviet Withdrawal from Afghanistan*, 45.
22. "Afghanistan: Recent Developments" (Note for the Secretary General), October 28, 1988, SML, Pérez de Cuéllar Papers, Box 10, Folder 110.
23. Record of meeting between Benon Sevan and Michael Armacost, November 14, 1988, SML, Pérez de Cuéllar Papers, Box 10, Folder 103.
24. Author's interview with Pavel Palazhchenko, Moscow, March 20, 2008; Author's interview with Leonid Shebarshin, Moscow, March 19, 2008.
25. Notes on a meeting between US undersecretary of state Michael Armacost and Benon Sevan, November 14, 1988, SML, Pérez de Cuéllar Papers, Box 10, Folder 103.
26. Author's interview with ambassador Yulii Vorontsov, Moscow, September 11, 2007.
27. Ibid. See also notes on Gorbachev and Pérez de Cuéllar meeting, December 7, 1988, and Talking Points prepared for Pérez de Cuéllar; SML, Pérez de Cuéllar Papers, Box 10, Folder 103.
28. Record of conversation between Vladimir Petrovsky, deputy foreign minister of the USSR, and Pérez de Cuéllar, October 21, 1988, SML, Pérez de Cuéllar Papers, Box 10, Folder 102.
29. This confusion is reflected in a conversation between Soviet deputy foreign

minister Petrovsky and Pérez de Cuéllar on October 21, 1988. Although Moscow seemed to be throwing all its weight behind Najibullah, Petrovsky "indicated that [Najibullah] would leave the scene." SML, Pérez de Cuéllar Papers, Box 10, Folder 102.

30. Sotskov, *Dolg i Sovest*, 1:101–108. This argument was also presented to Shevardnadze by Muhammad Gulabzoi, the minister of internal affairs. Handwritten notes taken at Shevardnadze's meeting with Gulabzoi and Tanai, September 1988, provided to the author by Dr. Antonio Giustozzi, London School of Economics.
31. Slinkin, *Afganistan vremen Taraki i Amina*, 295–296.
32. Ibid., 211–212.
33. Sotskov, *Dolg i Sovest*, 1:113.
34. Ibid., 113–115.
35. Ibid., 116.
36. Ibid., 118–119.
37. Handwritten notes taken at Shevardnadze's meeting with Gulabzoi and Tanai, September 1988, provided to the author by Dr. Antonio Giustozzi, London School of Economics.
38. Author's interview with Leonid Shebarshin, Moscow, March 19, 2008.
39. Sotskov, *Dolg i Sovest*, 120.
40. Varennikov, *Nepotovrimoe*, 376.
41. Ibid., 369–370.
42. Liakhovskii and Nekrasov, *Grazhdanin, Politik, Voin*, 169.
43. V. I. Varennikov, Memorandum to the USSR minister of defense, D. T. Yazov, August 1988, in Liakhovskii, *Tragediia i doblest'*, 656.
44. Sotskov, *Dolg i Sovest*, 111–113.
45. Ibid., 548–550.
46. Letter to Massoud, in Liakhovskii and Nekrasov, *Grazhdanin, Politik, Voin*, 193–194.
47. Ibid., 195–202.
48. Sotskov, *Dolg i Sovest*, 129–130.
49. Politburo meeting, September 18, 1988, GFA PB 1988.
50. Ibid. At this meeting, Shevardnadze also supported talks with Ahmad Shah Massoud. Later, as we will see, he became a forceful proponent of an attack on the Tajik leader.
51. Liakhovskii, *Tragediia i doblest'*, 669.
52. Kornienko, *Kholodnaia Voina*, 260.
53. Ibid., 261.
54. This version of events was also confirmed by Andrei Urnov, who worked on Afghan policy in the International Department between 1986 and 1989. Au-

thor's interview with Andrei Urnov, March 25, 2008. Yet there was another element to this story as well: the rivalry between Shevardnadze and Kornienko. Shevardnadze believed that Gorbachev was using Kornienko to "balance" Shevardnadze's influence in foreign policy. By 1988, Shevardnadze saw Kornienko as a potential opponent who would have to be removed. Author's interview with Shevardnadze, Tbilisi, Georgia, May 9, 2008.

55. Third conversation of M. S. Gorbachev and R. Gandhi (India), Delhi, November 19, 1988, NSA READD/RADD, Box 9.

56. Indeed, there is evidence that leaders of foreign communist parties were making their concerns about the Soviet withdrawal known to Moscow. A paper on Vietnamese foreign policy from IMEMO, a bastion of the New Thinking, noted that while the Vietnamese leadership saw the invasion of Afghanistan as a mistake, they were very unhappy with the way the Soviet Union had gone about the withdrawal. The paper, submitted to the Central Committee on August 30, 1988, noted that the Vietnamese were still dealing with a difficult situation in Cambodia and viewed the Soviet withdrawal in that light: "The Vietnamese are making it known that having taken the path of settling the Afghan problem, the Soviet Union has made excessive concessions to the opposing side. Soviet troops are being withdrawn hastily, despite the unfavorable development of events within Afghanistan, the unceasing aid of the USA and Pakistan." "Regarding the Foreign Policy of Vietnam," IMEMO Policy Paper, submitted to the CC CPSU August 30, 1988. IMEMO Archive.

57. Savranskaya, "Voenno-Politicheskie Aspekty Okonchaniia Kholodnoi Voiny, 62–64; Kornienko and Akhromeev, *Glazami Marshala i Diplomata*, 131–133.

58. Thus, in his conversation with Rajiv Gandhi on November 19, 1988, Gorbachev was also demonstrating that his control over the military had been restored.

59. Rogers, *Soviet Withdrawal from Afghanistan*, 144.

60. Record of conversation of M. S. Gorbachev with the president of Afghanistan and Najibullah, June 13, 1988, NSA READD/RADD, Box 9. Najibullah also suggested that the Soviet Union, Afghanistan, and India should launch a war against Pakistan, though it seems that this idea was never taken seriously. See Cherniaev Diary, June 19, 1988, GFA CD 1988.

61. For a fuller account of the military's relations with Massoud, see Liakhovskii, *Tragedia i doblest'*, 630–688; and Liakhovskii and Nekrasov, *Grazhdanin, Politik, Voin*.

62. Liakhovskii, *Tragediia i doblest'*, 651–652.

63. Politburo meeting, September 18, 1988, GF PB 1988.

64. Liakhovskii, *Tragediia i doblest'*, 616; see also Rogers, *The Soviet Withdrawal from Afghanistan*, 45.

65. Gromov, *Ogranichennyi Kontingent*, 324–325.
66. A. S. Cherniaev, Memorandum (for special 1606 from Kabul), October 26, 1988, GFA 1553.
67. Kriuchkov, *Lichnoe Delo*, 257.
68. Third conversation of M. S. Gorbachev and R. Gandhi (India), Delhi, November 19, 1988, NSA READD/RADD, Box 9. For a summary of Soviet "hints" that the withdrawal might be suspended indefinitely, hints that were dropped throughout November 1988, see "Afghanistan: Recent Developments" (Note for the SG), December 2, 1988, SML, Pérez de Cuéllar Papers, Box 10, Folder 110.
69. Politburo meeting, December 28, 1988, GFA PB 1988, 527.
70. Memorandum of conversation between Najibullah and Shevardnadze, Kabul, January 13, 1989, in Liakhovskii, *Tragediia i doblest'*, 670.
71. Unfortunately, there is no record available of how this decision was made. But Vorotnikov's diary does contain part of a January 13 Politburo meeting where Shevardnadze talked about an imminent economic blockade of Kabul. According to Vorotnikov's notes, Gorbachev said: "We must not leave the RA to its fate. Work. Think about propaganda. But first we leave, and then we act through the UN, the Security Council, and others" (see Vorotnikov, *A Bylo Eto Tak*, 280). It is difficult to evaluate Gorbachev's attitude from this fragment. On the one hand, he seems to be in favor of getting out and of using diplomacy only to protect the RA; on the other hand, he clearly emphasizes that the USSR must take responsibility for the RA. Yet because the operation was approved, we must assume that Shevardnadze and Kriuchkov convinced Gorbachev that it was necessary.
72. Bogdanov, *Afganskaia Voina*, 296–298.
73. Author's interview with General Aleksandr Liakhovskii, Moscow, July 3, 2006. Author's interview with Yulii Vorontsov, Moscow, September 11, 2007.
74. Sotskov, *Dolg i Sovest*, 531.
75. Memorandum to Gorbachev, October 26, 1988, GFA 1553.
76. G. Shakhnazarov, Memorandum to M. S. Gorbachev regarding Najibullah and Hekmatyar, December 16, 1988, GFA 18188.
77. Zagladin memorandum on Afghanistan, January 1989, GFA 7178.
78. "On Measures in Connection with the Upcoming Withdrawal of Soviet Troops from Afghanistan," January 23, 1989, in Allan et al., *Sowjetische Geheimdokumente*, 462–482.
79. Bogdanov, *Afganskaia Voina*, 294–295.
80. Author's interview with General Aleksandr Liakhovskii, Moscow, July 3, 2006.
81. Cherniaev, *Sovmestnyi iskhod*, 781-783.
82. Politburo meeting, January 24, 1989, GFA PB 1989, 48.
83. Chebrikov was still a Central Committee member, but no longer KGB chief.

84. Politburo meeting, January 24, 1989, GFA PB 1989, 48.
85. Politburo protocol, January 24, 1989, in Allan et al., *Sowjetische Geheimdokumente*, 458–460.
86. US military analyst Lester Grau has judged it an "excellent model for disengagement from direct military involvement in support of an allied government in a counter-insurgency campaign." Grau, "Breaking Contact without Leaving Chaos," 260–261.
87. "Regarding the Completion of Withdrawal of Soviet Troops from the Republic of Afghanistan," CC CPSU memorandum, February 16, 1989, Volkogonov Papers, Reel 17, Box 26.
88. IMEMO Policy Paper submitted to the CC CPSU, August 30, 1988, IMEMO Archive.
89. Zubok, *A Failed Empire*, 314–317.
90. Kotkin, *Armageddon Averted*, 87–92.

7. *Soviet Policy Adrift*

1. The study of this period poses particular methodological problems. As Gorbachev maneuvered among various factions, decision making became increasingly chaotic. At the same time, his effort from 1990 onward to distance himself from the party and create a new governing apparatus meant that, in practice, decisions were increasingly made informally or at the ministerial level. I have attempted to reconstruct some of the decision making for this period through a reading of newspaper reports, memoirs, and the few archival documents available.
2. Gareev, *Moia posledniaia voina*, 127.
3. Coll, *Ghost Wars*, 192–193; Liakhovskii and Nekrasov, *Grazhdanin, Politik, Voin*, 217–219; Yousaf and Adkin, *Bear Trap*, 226–228.
4. Coll, 194. Cherniaev, *Sovmestnyi iskhod*, 785–786.
5. Cherniaev, *Sovmestnyi iskhod*, 785–786; Politburo notes, March 10, 1989, GFA PB 1989, 202. Cherniaev was not at the meeting, and the information in this diary entry comes second-hand from Iakovlev. This version agrees in general with the notes in the Gorbachev Foundation Archive, which were taken by Vadim Medvedev.
6. Liakhovskii, *Tragediia i doblest'*, 682.
7. Cherniaev's notes of Politburo meeting, March 23, 1989, in Cherniaev, *Afganskii vopros*, 50.
8. Coll, *Ghost Wars*, 193–194; Yousaf and Adkin, *Bear Trap*, 230–232; "After Jalalabad's Defense, Kabul Grows Confident," *New York Times*, April 30, 1989; Author's interview with Ahmad Muslem Hayat, London, June 2010.
9. "After Jalalabad's Defense," *New York Times*, April 30, 1989.
10. See, for example, "The Encircling Gloom," *The Guardian*, April 10, 1989.

11. Gareev, *Moia posledniaia voina*, 127.
12. Militias had been used since 1980, but played an increasingly important role in the latter half of the 1980s, particularly after the withdrawal. The militias' loyalty to the Kabul government hinged largely on the payments and resources provided to them. See Giustozzi, *War, Politics, and Society*, 198–231.
13. Gareev, *Moia posledniaia voina*, 120.
14. Ibid., 127.
15. Ibid., 92.
16. Ibid. Gareev shares the frustration of other Soviet military officers who felt that their voices were often silenced by Kriuchkov. See Gareev, *Afganskaia Strada*, 276–277.
17. Gareev, *Moia posledniaia voina*, 132.
18. Ibid., 132–133.
19. Ibid., 134.
20. Although, not surprisingly, there was some suspicion among Soviet diplomats and officers in Kabul that the whole plot was hatched by the KGB. Gareev, *Moia posledniaia voina*, 135. Considering Kriuchkov's support for Najibullah both before and after the coup, it is most unlikely that the KGB would have been supporting Tanai's bid for power.
21. Coll, *Ghost Wars*, 212; Gareev, *Moia posledniaia voina*, 134–135.
22. Coll, *Ghost Wars*, 212; Rubin, *Fragmentation of Afghanistan*, 253.
23. Author's interview with Ahmed Muslem Hayat, London, June 2010.
24. Politburo meeting, March 7, 1990, GFA PB 1990, 169.
25. Shebarshin, *Iz Zhizni Nachal'nika Razvedki*, 24–25.
26. Tarzi, "Politics of the Afghan Resistance Movement."
27. Hussain, *Pakistan and the Emergence of Islamic Militancy*, 148–151.
28. Author's interview with Leonid Shebarshin, Moscow, March 19, 2008; Gareev, *Moia posledniaia voina*, 107–108.
29. "Regarding Talks in Kabul and Our Potential Further Steps . . . ," August 11, 1989, in Allan et al., *Sowjetische Geheimdokumente*, 692.
30. Cherniaev memorandum to Gorbachev, GFA Fond 2, Opis 1, Document 8242.
31. Author's interview with Nikolai Kozyrev, Moscow, November 14, 2008.
32. Conversation with author at American Historical Association Annual Meeting, San Diego, California, January 2010.
33. Lieven, *America Right or Wrong*, x.
34. Meeting between UN Secretary General and Zain Noorani, foreign minister of Pakistan, April 14, 1988, SML, Pérez de Cuéllar Papers, Box 10, Folder 102. Javier Pérez de Cuéllar's resistance to a UN role was broken by a UN resolution, and he ended up playing a direct role, while Diego Cordovez left the UN to serve as foreign minister for his own country, Ecuador. See Pérez de Cuéllar

to Cordovez, November 11, 1988, UN Secretary General's Files S-1031-60-25; and Pérez de Cuéllar, *Pilgrimage for Peace*, 196–197.

35. Pérez de Cuéllar, *Pilgrimage for Peace*, 202.
36. Ibid. See also a copy of the proposal handed to Pérez de Cuéllar, March 15, 1989, and the Note for File on US and Pakistani responses to these proposals (undated but after March 16, 1989), in SML, Pérez de Cuéllar Papers, Box 10, Folder 104.
37. Ekedahl and Goodman, *The Wars of Eduard Shevardnadze*, 193.
38. Note of the Secretary General's Meeting with the First Deputy Representative of the USSR, March 22, 1989, SML, Pérez de Cuéllar Papers, Box 10, Folder 109.
39. See, for example, Shevardnadze's note to Pérez de Cuéllar, February 17, 1989, and Vorontsov to Pérez de Cuéllar, November 6, 1990, SML, Pérez de Cuéllar Papers, Box 10, Folder 109.
40. Ekedahl and Goodman, *The Wars of Eduard Shevardnadze*, 193; Beschloss and Talbott, *At the Highest Levels*, 62; Baker, *The Politics of Diplomacy*, 74. Shevardnadze's hint that the Soviet Union might drop its insistence on keeping Najibullah in a coalition government was "off the record," and was not mentioned in a memorandum prepared by Soviet officials for the UN Secretary General, nor does Baker mention it in his memoirs. "USSR-USA Talks on Afghanistan," May 17, 1989, SML, Pérez de Cuéllar Papers, Box 10, Folder 104.
41. In September, Shevardnadze traveled to Baker's ranch at Jackson Hole, Wyoming, for two days of talks on arms control and other bilateral issues. Although Baker and Shevardnadze spoke about the Afghan problem, they could agree only on the need for a "political settlement on the basis of national reconciliation, and for a transitional government paving the way for the creation of a non-aligned Afghanistan." "Afghanistan: US Goes Cool on Guerillas," *The Guardian*, October 6, 1989.
42. Indeed, mid-level and senior officials had begun reevaluating their policies toward support for the *mujahadeen*. In the fall of 1989, Peter Tomsen, who had been appointed as ambassador to the Afghan resistance, led an interagency working group to reevaluate US policy. They decided on a new approach: pressure on Najibullah would continue, but the United States would work to form a moderate government to take his place. Coll, *Ghost Wars*, 180–184, 205–207.
43. Pérez de Cuéllar, *Pilgrimage for Peace*, 203; Khan, *Untying the Afghan Knot*, 306–307. This was also evident when Pakistani prime minister Benazir Bhutto, in a June meeting with Pérez de Cuéllar, noted that "the situation around Jalalabad had made everybody think in a different perspective. . . . Pakistan was pursuing a search for a political settlement to the problem of Afghanistan."

Notes on the meeting between the Secretary General and the prime minister of Pakistan, June 9, 1989, SML, Pérez de Cuéllar, Box 10, Folder 105.

44. "Afghanistan: US Goes Cool on Guerillas," *The Guardian*, October 6, 1989. See also Bush and Scowcroft, *A World Transformed*, 134–135.
45. Pérez de Cuéllar, *Pilgrimage for Peace*, 203–204.
46. "Soviets Reassert Policy on Keeping Najibullah," *Washington Times*, November 2, 1989, A7. Associated Press, "Soviet Support for Najibullah Blocked Political Headway at Malta Summit," December 6, 1989. Gorbachev to Pérez de Cuéllar, December 3, 1989, GFA.
47. Cherniaev Diary entry, December 10, 1989, GFA CD 1989.
48. "US Divided on Soviet Stand," *New York Times*, December 16, 1989, 1.
49. Beschloss and Talbott, *At the Highest Levels*, 180. See also "Moscow Spells Out Afghan Plan," *The Independent*, February 21, 1990, 9. Moscow's position, as laid out to the UN Secretary General, was that the US insistence on removing Najibullah during the transition period was unacceptable. However, Moscow was ready to accept the results of elections, as were Afghan leaders. Untitled memorandum on Baker-Shevardnadze talks, February 14, 1990, SML, Pérez de Cuéllar Papers, Box 10, Folder 106.
50. "Najibullah 'Can Remain,'" *The Guardian*, February 6, 1990.
51. Beschloss and Talbott, *At the Highest Levels*, 180
52. "Soviet Spokesman on Baker-Shevardnadze Talks in Moscow," *TASS*, February 8, 1990.
53. "Afghan Leader Urges UN-Monitored Elections," *Washington Post*, January 25, 1990, A30.
54. "Soviets Offer Proposal for Afghan Settlement," *Washington Post*, February 15, 1990, A45; Beschloss and Talbott, *At the Highest Levels*, 243.
55. "Superpowers Plan Afghan Arms Freeze," *The Independent*, April 5, 1990, 14; "US and Soviets on New Tack in Effort to End Afghan War," *New York Times*, May 3, 1990, A1; "Baker Notes Gains on Afghan Accord," *Washington Post*, June 14, 1990, A34. Note on Afghanistan to the Secretary General (undated, but after May 23, 1990), SML, Pérez de Cuéllar Papers, Box 10, Folder 106.
56. Note of the Secretary General's luncheon with George Bush, June 4, 1990, SML, Pérez de Cuéllar Papers, Box 10, Folder 106.
57. "Shevardnadze, Khomeini, Meet in Tehran," *Washington Post*, February 27, 1989.
58. "Regarding Talk in Kabul and Our Potential Further Steps . . . ," August 11, 1989, in Allan et al., *Sowjetische Geheimdokumente*, 686.
59. "Iran Halts Arms Flow to Afghan Shiites," *Toronto Star*, October 1, 1989, H8.
60. See the IMEMO paper prepared for the Ministry of Foreign Affairs, "International Aspects of a Settlement in Afghanistan," July 23, 1990, IMEMO Archive;

and "Iran's Attitude to the Settlement of the Afghan problem," GARF Fond 10026, Opis 4, Delo 2868.

61. Note for the Secretary General following talks regarding Afghanistan held with senior Soviet officials, July 10, 1990. The consensus non-paper, sent to the foreign ministers of Iran, Pakistan, the United States, and the USSR, outlined a number points which were necessary for securing a settlement, including the cessation of arms supplies by all sides and a "credible and impartial transition mechanism," with specifics on the latter point deliberately left out. "Elements for an International Consensus," July 11, 1990, SML, Pérez de Cuéllar Papers, Box 10, Folder 107.
62. "Record of Conversation between Gorbachev and Najibullah," August 23, 1990, NSA, READ/RADD, Box 9.
63. Beschloss and Talbott, *At the Highest Levels*, 243; Ekedahl and Goodman, *The Wars of Eduard Shevardnadze*, 193; "Superpowers Plan Afghan Arms Freeze," *The Independent*, April 5, 1990, 14.
64. Pankin, *Last Hundred Days*, 117–118; Author's interview with Nikolai Kozyrev, Moscow, November 14, 2008.
65. Grachev, *Gorbachev's Gamble*, 103.
66. Meeting chaired by Iakovlev, February 22, 1988, NSA, READ/RADD, Box 9, p. 17.
67. Politburo Meeting, April 18, 1988, GFA PB 1988, 211–215. At this meeting Gorbachev also nominated Iakovlev to the Afghan Commission, presumably to handle questions of propaganda, and possibly also to have a close ally in that body.
68. CC CPSU Letter on Afghanistan, May 10, 1988, NSA Afghanistan, Document 21, online at www.gwu.edu/~nsarchiv/NSAEBB/NSAEBB57/soviet.html (accessed December 27, 2010). Translated by Svetlana Savranskaya.
69. Notes from the 19th Party Congress, June 29, 1988, GFA.
70. Politburo meeting, January 24, 1989, GFA PB 1989, 60.
71. "Sobliudat Soglashenie," *Pravda*, February 18, 1989.
72. Zagladin Memorandum on Afghanistan, February 29, 1989, GFA Fond 3, Opis 1, Document 7192.
73. Ibid.
74. Politburo meeting, January 24, 1989, GFA PB 1989, 60; and in Cherniaev, *Afganskii vopros*, 48.
75. Interview by the author and Sergei Radchenko with Georgii Arbatov, Moscow, March 24, 2008.
76. Beschloss and Talbott, *At the Highest Levels*, 123.
77. See, for example, Varennikov's interview in *Ogonek* with journalist Artem Borovik in 1989; and the journalist N. Topuridze's interview with General Kim

Tsagolov, "Afghanistan: Voinu proigrali politiki," in Topuridze, *Argumenty i Fakty*. Tsagolov went so far as to say: "The decision to withdraw our troops is a manifestation of great civil courage on our part as well as that of the Afghan leadership. It is precisely that [courage] which was lacking in our previous leadership."

78. Most notable is the work of Artem Borovik, and of the journalists David Gai and Vladimir Snegirev. Gai and Snegirev published a series of articles in *Vechernia Moskva* (Evening Moscow) in the summer and fall of 1989, and eventually a book, *Vtorzhenie*. See, for example, Gai, "Afganistan: Kak Eto Bylo."
79. Such was the case with a piece prepared by Aleksandr Bovin on the situation in Afghanistan in December 1988. Bovin's report, based on his visit to Afghanistan, noted that "the withdrawal of Soviet troops is not accompanied by increased stabilization in the country" and went on to highlight the continued divisions within the PDPA and other problems. Though the piece was excised from the December 11 broadcast, Bovin saw to its publication in *Argumenty i Fakty* less than a week later. Bovin, "Pis'mo v redaktsiu."
80. Liakhovskii, *Tragediia i doblest'*, 750–751; Odom, *Collapse of the Soviet Military*, 284–285.
81. Galeotti, *Afghanistan: The Soviet Union's Last War*, 79–83.
82. The assertion of sovereignty and in some cases independence by the governments of the Soviet republics contributed to Moscow's budgetary crisis, as republic leaders curtailed taxes (as well as agricultural products) that they had formerly sent to Moscow. See Gaidar, *Collapse of an Empire*, 228–242.
83. Rubin, *Fragmentation of Afghanistan*, 171.
84. "Gardez Victory: Soviet Message of Support Revives Kabul Regime," *Agence France-Presse*, October 14, 1991.
85. Although one cannot be sure, there seems to be a strong hint of irony in the letter. Najibullah almost certainly knew the position that Iakovlev was taking on support for Afghanistan. First, Najibullah may have had such information from Kriuchkov, who had a very acrimonious relationship with Iakovlev that continued to play out long after the Soviet collapse. Second, Iakovlev's role as one of the "most liberal" people in the leadership was a matter of public comment in the Soviet and foreign press. Thus, it is hard to imagine that the following is written without a trace of irony: "I am well aware of your attention and tireless efforts directed at providing aid to the long-suffering Afghan people, of the constant support the Soviet Union provides to the Republic of Afghanistan. . . . Our government and our people highly value your persistent efforts and are grateful to your for this." Najibullah to Iakovlev, [translated from Dari], December 28, 1989, GARF F. A-10063, Opis 2, Delo 56, 4–5.

86. "Record of Conversation between Gorbachev and Najibullah," August 23, 1990, NSA, READ/RADD, Box 9.
87. Garthoff, *The Great Transition*, 457–460, 470–483; Pikhoia, *Sovetskii Soiuz*, 570–632.
88. Author's interview with Leonid Shebarshin, Moscow, March 19, 2008.
89. Brown, *Gorbachev Factor*, 270–271. It is possible that fears of a military coup, rumors of which had circulated from March to September 1990, also motivated Gorbachev to rely more closely on Kriuchkov. Odom, *Collapse of the Soviet Military*, 339–341.
90. The failed coup also led to the arrest of Varennikov and the suicide of Akhromeev.
91. Brown, *Gorbachev Factor*, 277–279.
92. Note for the Secretary General on Afghanistan, April 12, 1991, SML, Pérez de Cuéllar Papers, Box 10, Folder 107.
93. Pankin, *Last Hundred Days*, 117–118. Negotiations on a mutual cutoff date had been going on throughout 1991. Moscow also insisted that a cutoff could not take place without a cease-fire. Notes on meeting between Sevan, Picco, and Nikolai Kozyrev, March 13, 1991, SML, Pérez de Cuéllar Papers, Box 10, Folder 107.
94. "US, Soviets Agree to Halt Arms to Combatants in Afghanistan," *Washington Post*, September 14, 1991, 1.
95. Pankin, *Last Hundred Days*, 117–118. Nikolai Kozyrev confirmed in an interview with me that Kriuchkov's influence in this period was crucial for maintaining Soviet deliveries to Najibullah. Author's interview with Nikolai Kozyrev, Moscow, November 14, 2008.
96. "US, Soviets Agree to Halt Arms to Combatants in Afghanistan," *Washington Post*, September 14, 1991, 1.
97. Pankin, *Last Hundred Days*, 118.
98. See, for example, "No Friends for Najibullah," *The Economist*, October 12, 1991, 34.
99. "Gardez Victory, Soviet Message of Support Revives Kabul Regime," *Agence France-Presse*, October 14, 1991. This position was reiterated by a Soviet Foreign Ministry spokesman. See "Soviet Approach to the Settlement of the Afghan Issue," *TASS*, October 14, 1991.
100. "Afghan Rebels Meet Rutskoy," *Agence France-Presse*, November 11, 1991.
101. Maley, *Afghanistan Wars*, 187.
102. Kniazev, *Istoriia Afganskoi Voiny*, 56. See also "Boris Yeltsin's First 100 Days," *Heritage Foundation Backgrounder*, 869, November 27, 1991, available at www.heritage.org/Research/RussiaandEurasia/bg869.cfm (accessed October 17, 2008).

103. Prior to the August putsch, the Soviet Ministry of Foreign Affairs had not allowed Russian politicians to play any role on Afghanistan. In the fall of 1991, they found this much harder because the Russian government was increasingly treated as an equal by foreign leaders and also had easier access to financial resources. Now it was the Russian MFA that sought to sideline Soviet diplomats. I. Adrionov to Rutskoi, Memorandum "On Talks with the Delegation of Afghan *Mujahadeen*," undated, September or October 1991, GARF, Fond 10026, Opis 4, Delo 2840, 30–34.
104. Associated Press, "Rebels Say Russia Supports Their Struggle against Kabul Government," November 11, 1991; "Soviet-Afghan Talks in Moscow," *TASS*, November 12, 1991; "Afghan Leader Feels Moscow Draught," *The Guardian*, November 19, 1991; Pankin, *Last Hundred Days*, 120–121.
105. Summary of discussions at Malta presented to Pérez de Cuéllar by ambassador Thomas R. Pickering, December 7, 1989, SML, Pérez de Cuéllar Papers, Box 10, Folder 109. Bush focused in particular on Nicaragua and Cuba.
106. This includes Aleksandr Rutskoi, a fighter pilot shot down in Afghanistan and briefly held as a POW. Yeltsin picked him as vice president in part to deflect potential criticism from the security forces. Although Rutskoi proved to be an unreliable ally in most ways, he never challenged Yeltsin on his Afghan policy. Indeed, as we saw earlier, Rutskoi took the lead in meeting with *mujahadeen* leaders and promising to end support for Najibullah.

Conclusion

1. Rubin, *Fragmentation of Afghanistan*, 71.
2. Author's interview with Abdul Razak, Moscow, July 2010.
3. Rubin, *Fragmentation of Afghanistan*, 269–270.
4. Liakhovskii, *Tragediia i doblest'*, 702.
5. The story of Kabul's last weeks under Najibullah and the valiant effort of UN officials to arrange a transfer of power and avert an intra-*mujahadeen* civil war is told in Corwin, *Doomed in Afghanistan*.
6. Gareev, *Afganskaia Strada*, 318.
7. *Pravda*, April 13, 1993.
8. See, for example, the interview with Nikolai Kozyrev in *Moskovskii Komsomolets*, March 5, 2004.
9. Garthoff, *Great Transition*, 464–465, 470–471, 489–496.
10. Crile, *Charlie Wilson's War*, 518–523; Coll, *Ghost Wars*, 240–241, 262–265.
11. Quoted in Coll, *Ghost Wars*, 265.
12. Kux, *The United States and Pakistan*, 308–358; Hilali, *US-Pakistan Relationship*, 231–242.
13. Rubin, *Fragmentation of Afghanistan*, 265–275.
14. Ibid., 275–277.

15. Rashid, *Taliban: Islam, Oil, and the New Great Game,* 19–30.
16. Ibid., 32.
17. Liakhovskii and Nekrasov, *Grazhdanin, Politik, Voin,* 282.
18. Coll, *Ghost Wars;* Rashid, *Taliban: Islam, Oil, and the New Great Game,* 157–182; Magnus and Naby, *Afghanistan: Mullah, Marx, and Mujahid,* 196–206.
19. Jackson, *Russian Foreign Policy and the CIS,* 140–170; Liakhovskii and Nekrasov, *Grazhdanin, Politik Voin,* 294–301.
20. The exception, as of this writing, may be the war with Georgia in August 2008.
21. On the Horn of Africa, see Westad, "The Fall of Détente and the Turning Tides of History," in Westad, ed., *The Fall of Détente,* 3–33. On Hungary and Czechoslovakia, see Bekes, "The 1956 Hungarian Revolution and World Politics"; Zubok, *Failed Empire,* 115–119, 207–209; Pikhoia, *Sovetskii Soiuz,* 301–343.
22. Zubok, *Failed Empire,* 267.
23. See Ouimet, *The Rise and Fall of the Brezhnev Doctrine.* Looking at Soviet behavior in Eastern Europe in the 1970s and early 1980s, Ouimet persuasively shows that the Brezhnev Doctrine had largely been rejected by the time of the Polish crisis.
24. Politburo meeting, June 11, 1987, GFA PB 1987, 342.
25. Rubin, *Fragmentation of Afghanistan,* 164. Even technical and agricultural aid often had unintended consequences, and the aid effort has now begun to attract attention from Western analysts. See Robinson, "Russian Lessons."
26. Many contemporary observers did not see it this way, explaining the presence of Soviet advisers as a program of "Sovietization." One observer, arguing that Afghanistan was being "Sovietized" on the Central Asian model, wrote: "When Soviet leaders hint at a possible willingness to withdraw military forces, they say nothing about withdrawing their second army—the army of social and cultural transformation, spearheaded by the KGB—or dismantling the programs designed to accomplish this end." Amin, "The Sovietization of Afghanistan," 334.
27. The literature on Soviet aid efforts in the Third World is quite sparse. Although a number of works distill Soviet thinking about modernization and transitions to socialism in the Third World, there is no parallel to works such as Latham's *Modernization as Ideology,* Gilman's *Mandarins of the Future,* or Simpson's *Economists with Guns,* which are concerned with the ideas and practice of US aid to the Third World.
28. Notes from Politburo meeting, July 23, 1987, GFA PB 1987, 429.
29. The new consensus in the Politburo was that the PDPA would be only one of the political forces in power after Soviet troops left. Even Kriuchkov agreed that reconciliation would have to take place not around the PDPA, but with its participation. Gromyko, too, said that the PDPA should be one of the parties in

the government, but not the leading one. See Politburo meeting, May 22, 1987, GFA PB 1987, 319.

30. It is worth noting the perceptive assessment by RAND analyst Tad Daley: "The costs of staying in Afghanistan did not come to exceed the costs of leaving because of any dramatic changes in the tangible costs of the occupation. What changed instead was the new Soviet leadership's perception of the nature of those costs themselves." Daley, "Afghanistan and Gorbachev's Global Foreign Policy," 497. This is particularly accurate of the months prior to the signing of the Geneva Accords and immediately afterward—when the possibility of a major breakthrough in US-Soviet relations made a withdrawal, even one under less than ideal conditions, seem much more attractive than previously.
31. See Zubok, "Gorbachev and the End of the Cold War," 79–82, 92–93.
32. See Mendelson, "Internal Battles and External Wars," 327–360; and Mendelson, *Changing Course*.
33. As it did for Kornienko in the fall of 1988. Gorbachev was not completely deaf to the entreaties of his advisers who disagreed with Kriuchkov and Shevardnadze. For example, when Najibullah requested Soviet air support to defend Jalalabad against a major *mujahadeen* onslaught in March 1989, he was dissuaded despite the strong endorsement of such a move by Shevardnadze and Kriuchkov. See Cherniaev, *Afganskii vopros*, 49–50; Politburo notes, March 10, 1989 GFA PB 1989, 202.
34. Most famously, the military counterinsurgency manual overseen by US Army General David Petraeus. "Counterinsurgency," US Army, 2006. Online at www.fas.org/irp/doddir/army/fm3–24.pdf (accessed March 21, 2009). For a survey of the dilemmas faced by Great Powers fighting counterinsurgencies, see Edelstein, "Occupational Hazards," especially the section entitled "The Dilemma of Failing Occupations."
35. In the following section, I have relied on several general works on Vietnam, including Young, *The Vietnam Wars, 1945–1990*; Herring, *America's Longest War*; Gardner and Young, *Iraq and the Lessons of Vietnam*. I am also grateful to several scholars who provided me with papers prepared for a conference called "The Politics of Troop Withdrawal," held in June 2008 at the Miller Center, University of Virginia. These are Robert Jervis, "The Politics of Troop Withdrawal: Salted Peanuts, the Commitment Trap, and Buying Time"; and Robert J. McMahon, "The Politics, and Geopolitics, of American Troop Withdrawals from Vietnam, 1968–1972." All of these have now been published in a special edition of *Diplomatic History*.
36. In his study comparing the US involvement in Vietnam with the Soviet involvement in Afghanistan, Douglas Borer analyzes some of the political dilemmas involved, as well as the effect of the wars on the superpowers. His book contains some valuable insights, including this statement on the dilemma of

intervention in support of an unpopular regime: "We can now understand that superpower intervention in Vietnam and Afghanistan created an irreconcilable contradiction: without direct military support, the regimes in Saigon and Kabul could not survive; yet with superpower intervention the regimes undermined their chances of convincing their populations that they were legitimate governments." Borer, *Superpowers Defeated*, 197. Yet Borer greatly exaggerates the effect the war had on the Soviet Union, seeing it as one of the main causes of the USSR's collapse.

37. Westad, *Global Cold War*, 397.
38. For example, Arnold, *The Fateful Pebble*; Reuveny and Prakash, "The Afghanistan War and the Breakdown of the Soviet Union."
39. The war's most important effect on the disintegration of the Soviet military was that it led to widespread draft evasion. But other aspects of the Soviet military's decline in 1989–1991 were caused by additional factors, including rising nationalism in the constituent republics; the questioning of all Soviet institutions, including the military, that resulted from glasnost; and the collapse of the Soviet state. See Odom, *Collapse of the Soviet Military*, 247–251, 272–304.
40. "Afghanistan: New US Administration, New Directions," *ICG Asia Briefing* 89, March 2009.
41. See the discussion of Obama's options in a recent paper from the Council of Foreign Relations: Markey, "From AfPak to PakAf." The debates within the administration were detailed in the recent book by journalist Bob Woodward, *Obama's Wars*.
42. For more on this debate, see the roundtable with Scott Lucas, Andrew Johnson, Scott Smith, Marilyn Young, and Artemy Kalinovsky, *Neoamericanist*, Autumn 2009. See also Kalinovsky, "Afghanistan Is the New Afghanistan."
43. Woodward, *Obama's Wars*, 286.
44. Gromov and Rogozin, "Russian Advice on Afghanistan."
45. Weitz, "Russia Returns to Afghanistan."
46. Blackwill, *Plan B in Afghanistan*, 44.

A NOTE ON SOURCES

Researching this period in Soviet history poses a number of problems, the greatest of which is the limited access to primary materials. Certain aspects of decision making and policy implementation are virtually impossible to trace in the documentary record. I have overcome these difficulties by using all possible archival resources in Russia, supplementing those materials with published documents, memoir literature, oral history, and archival research in the Unites States. I have also made use of the numerous memoirs about general Afghan and Soviet politics during the 1970s and 1980s, and conducted interviews with some of the key figures in Soviet military, intelligence, diplomacy, and party leadership who helped to shape or carry out Afghan policy.

Certainly the best possible sources for understanding Soviet policy making are high-level memoranda and minutes of meetings that provide a record of the debates. Their selective availability, however, presents some problems. The official ones that were released as part of "Fond 89" (held at the Russian State Archive of Contemporary History, or RGANI; a "fond" is a collection of documents generated by a single entity) were selected in part to embarrass the Soviet Communist Party. Thus, with regard to Afghanistan, most of them focus on the decision to invade, not on the conduct of the war or on international diplomacy in the 1980–1991 period. This gap is partly filled by the collection at the Gorbachev Foundation Archive (GFA), compiled by Gorbachev's associates (including Anatolii Cherniaev, his key foreign-policy aide), which includes a collection of Politburo documents, many of them minutes of key meetings. These are handwritten notes, not a formal record. Yet wherever I have been able to cross-reference these with the documents in state archives, I have found them to be notably consistent. The materials in the GFA are particularly useful for tracing debates at the Politburo level during the Gorbachev period, at least through 1990. The various attempts to reform the decision-making apparatus, which decreased the importance of the Politburo, and the general disintegration of the state make these notes much less useful for the 1990–1991 period.

While one might expect these and other documents in the GFA to have been selected in a way that sheds a positive light on Gorbachev, the picture they present is far from one-sided. The bias they represent is primarily that of availability: the documents are those that came across the desk of one of the aides or that were drafted by those aides. They do not give anywhere near a complete picture of the conduct of foreign policy in a way that would be possible if we had access to materials from the Foreign Ministry, the International Department of the CC CPSU (Central Committee of the Communist Party of the Soviet Union), the military, and the intelligence agencies.

To compensate for these lacunae, I have made use of a number of archives that hold formerly declassified documents, or materials related to my topic, that help to illuminate the context in which decisions on Afghanistan were made. There are many documents at the National Security Archive in Washington, DC, retrieved from several Russian archives during the 1990s, when access was less restricted than it currently is. Many of these are records of conversations between representatives or leaders of the USSR and those of foreign countries, including Afghanistan. I have also made use of records at the State Archive of the Russian Federation (GARF), including the papers of Aleksandr Iakovlev, a leading reformer and member of the Afghan Commission of the Politburo under Gorbachev. GARF also holds the files of the Council of Religious Affairs, the body that regulated religious practice throughout the USSR and also tracked religious sentiment in the country, reporting its findings to the Central Committee.

Some archives remained closed. Intelligence and military archives are generally difficult to access, particularly for a period as recent as the one in question. Some documents from the military have appeared in the various memoirs published by officers who served in Afghanistan. Other documents were made available to me by General Aleksandr Liakhovskii and Dr. Antonio Giustozzi; I have specified in the notes those instances where I made use of these materials. At the Archive of the Foreign Ministry (AVPRF), I was allowed to see some of the files from the Kabul Embassy, but nothing above the level of press clippings and some largely irrelevant correspondence.

Translations are my own unless otherwise noted.

BIBLIOGRAPHY

Archives

GORBACHEV FOUNDATION, MOSCOW, RUSSIA

Anatolii Cherniaev Diaries
Anatolii Cherniaev Papers
Politburo Records, 1985–1991
Papers of Gorbachev Associates

RGANI: STATE ARCHIVE FOR CONTEMPORARY HISTORY, MOSCOW

Fond 89, Declassified documents, 1922–1991
Fond 5, CC CPSU Papers

GARF: STATE ARCHIVE OF THE RUSSIAN FEDERATION, MOSCOW

Fond 6991, Council of Religious Affairs files
Fond 10026, Papers of the Congress of People's Deputies of the Russian Federation
Fond 10063, Papers of Aleksandr Nikolaevich Iakovlev
Fond 9540, Papers of the Soviet Committee for Solidarity with Countries of Asia and Africa (SKSSAA)

RGAE: RUSSIAN STATE ARCHIVE OF THE ECONOMY

Fond 413, Ministry of Trade

NATIONAL SECURITY ARCHIVE, WASHINGTON, DC

Afghanistan Collection, Documents collected from US and Russian Archives
End of the Cold War Collection, Documents collected from Russian Archives
READ/RADD Collection, Documents collected from Russian Archives

REAGAN LIBRARY, SIMI VALLEY, CALIFORNIA

William P. Clark Papers
Head of State File
Jack Matlock Papers
National Security Council Papers

UNITED NATIONS ARCHIVES, NEW YORK, NY

Files of the UN Secretary General
Files of the Undersecretary General for Special Political Affairs

STERLING MEMORIAL LIBRARY, YALE UNIVERSITY, NEW HAVEN, CT

Papers of UN Secretary General Pérez de Cuéllar

LIBRARY OF CONGRESS MANUSCRIPT DIVISION, WASHINGTON, DC

Dmitrii Volkogonov Papers

Published Materials

Abbas, Hassan. *Pakistan's Drift into Extremism.* Armonk, NY: M. E. Sharpe, 2005.

Adamec, Ludwig. *Historical Dictionary of Afghanistan.* Lanham, MD: Scarecrow Press, 1997.

Adamsky, Dima. *The Culture of Military Innovation: The Impact of Cultural Factors on the Revolution in Military Affairs in Russia, the US, and Israel.* Stanford: Stanford University Press, 2010.

Allan, Pierre et al., ed. *Sowjetische Geheimdokumente zum Afghanistankrieg 1978–1991.* Zurich: Hochschulverlag, 1995.

Aleksandrov-Agentov, Andrei M. *Ot Kollontai do Gorbacheva.* Moscow: Mezhdunarodnie Otnoshenie, 1994.

Alexiev, Aleksandr. *Inside the Soviet Army in Afghanistan.* Santa Monica, CA: Rand Corporation, 1988.

Alexievich, Svetlana. *Zinky Boys: Soviet Voices from a Forgotten War.* London: Chatto and Windus, 1992.

Amin, A. Rasul. "The Sovietization of Afghanistan." In Roseanne Klass, ed. *Afghanistan: The Great Game Revisited,* ed. Roseanne Klass. New York: Freedom House, 1987.

Andrew, Christopher, and Vasilii Mitrokhin. *The World Was Going Our Way: The KGB and the Battle for the Third World.* New York: Basic Books, 2005.

Arbatov, Georgii. *Chelovek Sistemy: Nabliudenia i razmyshlenia ochevidtsa ee raspada*. Moscow: Vagrius, 2002.

Arnold, Anthony. *The Fateful Pebble: Afghanistan's Role in the Fall of the Soviet Empire*. Novato, CA: Presidio, 1993.

Atkin, Muriel. *The Subtlest Battle: Islam in Soviet Tajikistan*. Philadelphia: Foreign Policy Research Institute, 1989.

——— "The Islamic Revolution that Overthrew the Soviet State." *Contention* 2 (Winter 1993).

Avakov, Rachik M. "Novoe mishlenie i problema izuchenia razvivaiushikhsia stran" [New Thinking and the Problem of Studying Developing Countries]. *Mirovaia ekonomika i mezhdunarodnye otnosheniia* 11 (1987): 48–62.

Baker, James A. *The Politics of Diplomacy: Revolution, War and Peace, 1989–1992*. New York: G. P. Putnam's Sons, 1995.

Bakshi, Jyotsna. "Soviet Approach to the Problem of Afghanistan-Pakistan Settlement, 1976–1985." *India Quarterly* 50, nos. 1–2 (1994): 95–122.

Barfield, Thomas J. *Afghanistan: A Cultural and Political History*. Princeton: Princeton University Press, 2010.

Bearden, Milton, and James Risen. *The Main Enemy: The Inside Story of the CIA's Final Showdown with the KGB*. Novato, CA: Presidio, 2003.

Beissinger, Mark R. *Nationalist Mobilization and the Collapse of the Soviet State*. Cambridge: Cambridge University Press, 2002.

Bekes, Csaba. "The 1956 Hungarian Revolution and World Politics." CWIHP Working Paper 16. Washington: CWIHP, 1996.

Belonogov, Aleksandr Mikhailovich. *Na diplomaticheskoi avanscene: Zapiski postoiannogo predstavitelia SSSR pri OON* [On the Center Stage of Diplomacy: Notes of the Permanent Representative of the USSR at the UN]. Moscow: MGIMO-Universitet, 2009.

Bennett, Andrew. *Condemned to Repetition? The Rise, Fall, and Reprise of Soviet-Russian Military Interventionism, 1973–1996*. Cambridge, MA: MIT Press, 1999.

——— "The Guns That Didn't Smoke: Ideas and the Soviet Non-Use of Force in 1989." *Journal of Cold War Studies* 7, no. 2 (2005): 81–109.

Bennigsen, Aleksandre. "Afghanistan and the Muslims of the USSR." In Rosemary Klass, ed., *Afghanistan: The Great Game Revisited*. New York: Freedom House, 1987.

——— "Soviet Islamic Strategy after Afghanistan." In Alexandre Bennigsen et al., *Soviet Strategy and Islam*. Basingstoke, UK: Macmillan, 1989.

——— and Marie Broxup. *The Islamic Threat to the Soviet Union*. London: Croom Helm, 1993.

Beschloss, Michael R., and Strobe Talbott. *At the Highest Levels: The Inside Story of the End of the Cold War*. Boston: Little, Brown, 1993.

Best, Anthony, et al., eds. *The Twentieth Century: An International History.* London: Routledge, 2003.

Bisley, Nick. *The End of the Cold War and the Causes of Soviet Collapse.* New York: Palgrave Macmillan, 2004.

Blackwill, Robert. "Plan B in Afghanistan." *Foreign Affairs* 90, no. 1 (2011), 42–50.

Bogdanov, Vladimir Alekseevich. *Afganskaia voina: Vospominania.* Moscow: Sovetskii Pisatel, 2005.

Borer, Douglas A. "The Afghan War: Communism's First Domino." *War and Society* 12, no. 2 (1994): 127–144.

——— *Superpowers Defeated: Vietnam and Afghanistan Compared.* London: Frank Cass, 1999.

Borovik, Artem. *The Hidden War: A Russian Journalist's Account of the Soviet War in Afghanistan.* New York: Atlantic Monthly Press, 1990.

Bose, Sugata, and Ayesha Jalal. *Modern South Asia: History, Culture, Political Economy.* London: Routledge, 1998.

Bovin, Aleksandr. "Pis'mo v redaktsiu: glasnost na polovinu." *Argumenty i Fakty,* December 17, 1988.

Bradsher, Harry. *Afghan Communism and Soviet Intervention.* Oxford: Oxford University Press, 1999.

Braithwaite, Rodric. *Afgantsy: The Russians in Afghanistan,* 1979–1989. London: Profile Books, 2011.

Brown, Archie. *The Gorbachev Factor.* Oxford: Oxford University Press, 1996.

——— *Seven Years that Changed the World: Perestroika in Perspective.* Oxford: Oxford University Press, 2007.

——— "Perestroika and the End of the Cold War." *Cold War History* 7, no. 1 (2007): 1–17.

Brutents, Karen. *Tridsat' Let na Staroi Ploshadi.* Moscow: Mezhdunarodnye otnosheniia, 1998.

Brzezinski, Zbigniew. *Power and Principle: Memoirs of the National Security Adviser, 1977–1981.* London: Weidenfeld and Nicolson, 1983.

Bull, Hedley. "Intervention in the Third World." In Bull, ed., *Intervention in World Politics,* 135–156. Oxford: Clarendon Press, 1984.

Bush, George, and Brent Scowcroft. *A World Transformed.* New York: Knopf, 1998.

Caldwell, Dan. "US Domestic Politics and the Demise of Détente." In Westad, ed. *The Fall of Détente,* 95–117.

Calluther, Nick. "Damming Afghanistan: Modernization in a Buffer State." *Journal of American History* 89, no. 2 (2002): 512–537.

Campbell, K., and S. N. MacFarlane, eds. *Gorbachev's Third World Dilemmas.* New York: Routledge, 1989.

Carter, Jimmy. *Keeping Faith: Memoirs of a President.* London: Collins, 1982.

Chazov, Yevgeny. *Zdorovie i Vlast.* Moscow: Novosti, 1992.

Cherniaev, Anatolii S. *My Six Years with Gorbachev.* University Park: Pennsylvania State University Press, 2000.

——— "Afganskii Vopros" [The Afghan Question], *Svobodnaia Mysl'*, 11 (2002): 64–74; and 12 (2002): 39–50.

——— *Shest' let s Gorbachevym* [Six Years with Gorbachev]. Moscow: Progress, 1993.

——— *Sovmestnyi iskhod: Dnevnik dvukh epokh* [Joint Departure: A Diary of Two Eras]. Moscow: ROSSPEN, 2008.

Cohen, S. P. "South Asia after Afghanistan." *Problems of Communism* 34, no. 11 (Jan.–Feb. 1985): 18–31.

Coll, Steve. *Ghost Wars: The Secret History of the CIA, Afghanistan, and bin Laden, from the Soviet Invasion to September 10, 2001.* New York: Penguin, 2004.

Corwin, Phillip. *Doomed in Afghanistan: A UN Officer's Memoir of the Fall of Kabul and Najibullah's Failed Escape, 1992.* New Brunswick, NJ: Rutgers University Press, 2003.

Crile, George. *Charlie Wilson's War.* New York: Grove Press, 2003.

Cull, Nicholas J. *The Cold War and the United States Information Agency: American Propaganda and Public Diplomacy, 1945–1989.* Cambridge: Cambridge University Press, 2008.

Daley, Tad. "Afghanistan and Gorbachev's Global Foreign Policy." *Asian Survey* 29, no. 5 (1987): 496–513.

Daugherty, Leo J., III. "The Bear and the Scimitar: Soviet Central Asians and the War in Afghanistan, 1979–1989." *Journal of Slavic Military Studies* 8, no. 1 (1995): 73–96.

Del Pero, Mario. *The Eccentric Realist: Henry Kissinger and American Foreign Policy.* Ithaca: Cornell University Press, 2010.

DeNeufville, Peter B. "Ahmad Shah Massoud and the Genesis of the Nationalist Anti-Communist Movement in Northeastern Afghanistan, 1969–1979." Ph.D. diss., King's College, London, 2006.

Dobrynin, Anatolii. *In Confidence: Moscow's Ambassador to America's Six Cold War Presidents, 1962–1986.* New York: Random House, 1995. In Russian: *Sugubo Doveritel'no: Posol v Vashingtone pri Shesti Prezidentakh SShA.* Moscow: Avtor, 1996.

Dorronsoro, Gilles. *Revolution Unending: Afghanistan, 1979–Present.* New York: Columbia University Press, 2005.

Dubinin, Iuri Vladimirovich. "Na Postu Posla SSSR v Washingtone" [In the Post of the USSR's Ambassador in Washington]. *Novaia i Noveishaia Istoriia* 3 (2003): 77–101.

Duncan, Peter. *The Soviet Union and India.* London: Routledge, 1989.

Duncan, W. Raymond, and Carolyn McGiffert Ekedahl. *Moscow and the Third World under Gorbachev.* Boulder: Westview Press, 1990.

Edelstein, David. "Occupational Hazards: Why Military Occupations Succeed or Fail." *International Security* 29, no. 1 (2004): 49–91.

——— *Occupational Hazards: Success and Failure in Military Occupation.* Ithaca: Cornell University Press, 2008.

Egorychev, Nikolai. "Afganistan stoil nam 15 milliardov dollarov v god" [Afghanistan Cost Us 15 Billion Dollars a Year], interview with Evgenii Zhirnov, *Kommersant Vlast'*, 46, November 25, 2002.

Ekedahl, Carolyn McGiffert, and Melvin A. Goodman. *The Wars of Eduard Shevardnadze.* University Park: Pennsylvania State University Press, 1997.

Elvartanov, Ilya N. *Afganistan glazami ochevidtsa: Zapiski sovetnika* [Afghanistan through the Eyes of an Eyewitness: Notes of an Adviser]. Elista: Kalmykskoe Knizhnoe Izdatelstvo, 2000.

English, Robert D. *Russia and the Idea of the West: Gorbachev, Intellectuals, and the End of the Cold War.* New York: Columbia University Press, 2000.

Esposito, John L., ed. *Political Islam: Revolution, Radicalism or Reform?* Boulder: Lynne Riener, 1997.

Feifer, Gregory. *The Great Gamble: The Soviet War in Afghanistan.* New York: HarperCollins, 2009.

Fischer, Beth A. *The Reagan Reversal: Foreign Policy and the End of the Cold War.* Columbia: University of Missouri Press, 1997.

Fukuyama, Francis. *Gorbachev and the New Soviet Agenda in the Third World.* Santa Monica, CA: Rand Corporation, 1989.

Fursenko, Aleksandr, and Timothy Naftali. *Khrushchev's Cold War.* New York: Norton, 2006.

Gaddis, John Lewis. *Strategies of Containment: A Critical Appraisal of American National Security Policy during the Cold War.* New York: Oxford University Press, 2005.

Gai, David. "Afganistan, kak eto bylo: Voina glazami ee uchastnikov" [Afghanistan the Way It Was: The War through the Eyes of Its Participants]. *Vecherniaia Moskva,* October 30, 1989.

——— and Vladimir Snegirev. *Vtorzhenie: Neizvestnie Stranizi Neobiavlennoy Voyni.* Moscow: IKPA, 1991.

Gaidar, Yegor. *Collapse of an Empire: Lessons for Modern Russia.* Washington, DC: Brookings Institution Press, 2007.

——— *Afghanistan, the Soviet Union's Last War.* London: Frank Cass, 1995.

Gankovskii, Yurii. "Afghanistan: From Intervention to National Reconciliation." *Iranian Journal of International Affairs* 4, no. 1 (Spring 1992).

Gardner, Lloyd C., and Marilyn B. Young, eds. *Iraq and the Lessons of Vietnam.* New York: New Press, 2007.

Gareev, M. A. *Afganskaia strada (S Sovetskimi voiskami i bez nikh)*. Moscow: Insan, 1999.

——— *Moia posledniaia voina* [My Last War]. Moscow: Insan, 1996.

Garthoff, Raymond L. *Détente and Confrontation: American-Soviet Relations from Nixon to Reagan*. Rev. ed. Washington, DC: Brookings Institution, 1994.

——— *The Great Transition: American-Soviet Relations and the End of the Cold War*. Washington, DC: Brookings Institution, 1994.

Gates, Robert. *From the Shadows: The Ultimate Insider's Story of Five Presidents and How They Won the Cold War*. New York: Touchstone, 1997.

Gelman, Harry. *The Brezhnev Politburo and the Decline of Détente*. Ithaca: Cornell University Press, 1984.

Gibbs, David N. "Reassessing Soviet Motives for Invading Afghanistan: A Declassified History." *Critical Asian Studies* 38, no. 2 (2006): 239–263.

Gilman, Nils. *Mandarins of the Future: Modernization Theory in Cold War America*. Baltimore: John Hopkins University Press, 2007.

Giscard d'Estaing, Valéry. *Le Pouvoir et la vie*. Paris: Compagnie 12, 1991.

Giustozzi, Antonio. *War, Politics and Society in Afghanistan*. London: Hurst, 2000.

Goodman, Melvin A. *Gorbachev's Retreat: The Third World*. New York: Praeger, 1991.

Gorbachev, Mikhail. *Izbrannye rechi i stat'i*. Moscow: Politizdat, 1990.

——— *Memoirs*. London: Doubleday, 1996.

——— *Perestroika: New Thinking for Our Country and the World*. London: Fontana, 1988.

Grachev, A. S. *Gorbachev's Gamble: Soviet Foreign Policy and the End of the Cold War*. Cambridge, MA: Polity Press, 2008.

Grau, Lester W. "Breaking Contact without Leaving Chaos." *Journal of Slavic Military Studies* 20, no. 2, (2007): 235–261.

——— "The Soviet-Afghan War: A Superpower Mired in the Mountains." *Journal of Slavic Military Studies* 17, no. 1 (2004): 129–151.

——— and Ali Ahmad Jalali. "Forbidden Cross-Border Vendetta: SpetsNaz Strike into Pakistan during the Soviet-Afghan War." *Journal of Slavic Military Studies* 18, no. 4 (2005): 661–672.

——— and Ali Ahmad Jalali. "The Campaign for the Caves: The Battles of Zhawar in the Soviet-Afghan War." *Journal of Slavic Military Studies* 14, no. 3 (2001): 69–92.

——— and William A. Jorgensen. "Beaten by the Bugs: The Soviet-Afghan War Experience." *Military Review* 77, no. 6 (1997): 30–37.

——— and Mohammad Yahya Nawroz. "The Soviet Experience in Afghanistan." *Military Review* 75, no. 5 (1995): 17–27.

——— ed. *The Bear Went over the Mountain: Soviet Combat Tactics in Afghanistan*. Portland, OR: Frank Cass, 1998.

Gregorian, Vartan. *The Emergence of Modern Afghanistan: Politics of Reform and Modernization, 1880–1946*. Stanford, CA: Stanford University Press, 1969.

Gromov, B. V. *Ogranichennyi Kontingent* [The Limited Contingent]. Moscow: Progress, 1994.

——— and Dmitry Rogozin. "Russian Advice on Afghanistan," *New York Times*, January 11, 2010.

Haig, Alexander M., Jr. *Caveat: Realism, Reagan, and Foreign Policy*. New York: Macmillan, 1984.

Halliday, Fred. *Cold War, Third World: An Essay on Soviet-US Relations*. London: Hutchinson Radius, 1989.

——— "Islam and Soviet Foreign Policy." *Journal of Communist Studies and Transition Politics* 3, no. 1 (1987): 37–54.

——— "Soviet Foreign Policymaking and the Afghanistan War: From Second Mongolia to 'Bleeding Wound.'" *Review of International Studies* 25 (1999): 679–691.

Hammond, Thomas T. *Red Flag over Afghanistan: The Communist Coup, the Soviet Invasion, and the Consequences*. Boulder, CO: Westview Press, 1990.

Hanson, Philip. *The Rise and Fall of the Soviet Economy: An Economic History of the USSR from 1945*. London: Pearson, 2003.

Harrison, Selig, and Diego Cordovez. *Out of Afghanistan: The Inside Story of the Soviet Withdrawal*. New York: Oxford University Press, 1995.

Hartman, Andrew. "The Red Template: US Policy in Soviet-Occupied Afghanistan." *Third World Quarterly* 23, no. 3 (2002): 467–489.

Heinämaa, Anna, Maija Leppänen, and Yuri Yurchenko. *The Soldier's Story: Soviet Veterans Remember the Afghan War*. Trans. A. D. Haun. Berkeley: International and Area Studies, University of California at Berkeley, 1994.

Herring, George C. *America's Longest War: The United States and Vietnam, 1950–1975*. 3rd ed. New York: McGraw Hill, 1996.

Hershberg, James. "The War in Afghanistan and the Iran-Contra Affair: Missing Links." *Cold War History* 3, no. 3 (2003): 23–48.

Hilali, A. Z. *US-Pakistan Relationship: Soviet Invasion of Afghanistan*. Burlington, VT: Ashgate, 2005.

Hussain, Rizwan. *Pakistan and the Emergence of Islamic Militancy in Afghanistan*. Aldershot: Ashgate, 2005.

Jackson, Nicole J. *Russian Foreign Policy and the CIS: Theories, Debates and Actions*. London: Routledge, 2003.

Jallot, Nicolas. *Chevardnadzé: Le Renard blanc du Caucase*. Paris: Belfond, 2005.

Jervis, Robert. "The Politics of Troop Withdrawal: Salted Peanuts, the Commitment Trap, and Buying Time." *Diplomatic History* 34, no. 3 (2010): 507–516.

Kakar, Hasan. *Afghanistan: The Soviet Invasion and the Afghan Response, 1979–1982*. Berkeley: University of California Press, 1995.

Kalinovsky, Artemy M. "Afghanistan Is the New Afghanistan." *Foreign Policy* 4

(September 2009). Online at www.foreignpolicy.com/articles/2009/09/04/afghanistan_is_the_new_afghanistan (accessed January 3, 2011).

——— "Decision-Making and the Soviet War in Afghanistan, from Intervention to Withdrawal." *Journal of Cold War Studies* 11, no. 4 (2009): 46–73.

——— "Politics, Diplomacy and the Soviet Withdrawal from Afghanistan: From National Reconciliation to the Geneva Accords." *Cold War History* 8, no. 3 (August 2008): 381–404.

——— and Sergei Radchenko, eds. *The End of the Cold War and the Third World.* London: Routledge, 2011.

Kaufman, Barton I., and Scott Kaufman. *The Presidency of James Earl Carter.* 2nd rev. ed. Lawrence: University Press of Kansas, 2006.

Kevorkov, Viacheslav. *Tainyi Kanal* [The Secret Channel]. Moscow: Geia, 1997.

Khan, Riaz M. *Untying the Afghan Knot: Negotiating Soviet Withdrawal.* Durham, NC: Duke University Press, 1991.

Khristoforov, Vasilii. "Trudny put' k Zhenevskim Soglasheniam po Afganistanu." *Novaia i Noveishaia Istoriia* 5 (2008): 23–47.

Kirpichenko, Vadim. *Razvedka: Litsa i lichnosti* [Intelligence: Faces and Personalities]. Moscow: Geia, 1998.

Klass, Roseanne. "Afghanistan: The Accords." *Foreign Affairs* 66, no. 5 (1988): 922–945.

Kniazev, Aleksandr. *Istoriia Afganskoi Voiny 1990-h godov i prevrashenie Afganistana v istochnik ugroz Central'noy Azii.* Bishkek: Izd-vo Kyrgyzsko-Rossiĭskogo Slavîanskogo Universiteta, 2002.

Kolodziej, Edward A., and Roger E. Kanet, eds. *The Limits of Soviet Power in the Developing World.* Basingstoke: Macmillan, 1989.

Kolosovskii, Andrei I. "Regional'nye konflikty i globalaia bezopasnost'" [Regional Conflicts and Global Security]. *Mirovaia ekonomika i mezhdunarodnye otnosheniia* 6 (1988): 32–41.

Kornienko, G. M. *Kholodnaia voina: svidetel'stvo ee uchastnika* [The Cold War: A Participant's Account]. Moscow: Mezhdnuarodnie Otnosheniia, 2001.

——— "The Afghan Endeavor: Perplexities of the Military Incursion and Withdrawal." *Journal of South Asian and Middle Eastern Studies* 17, no. 2 (Winter 1994): 2–17.

——— and A. S. Akhromeev. *Glazami marshala i diplomata* [Through the Eyes of a Marshal and Diplomat]. Moscow: Mezhdunarodnie Otnosheniia, 1992.

Kotkin, Stephen. *Armageddon Averted: The Soviet Collapse, 1970–2000.* Oxford: Oxford University Press, 2001.

Kozovoi, Andrei. "La Ligne politique Soviétique en direction des Etats-Unis pendant la Guerre Fraiche, 1975–1985." *Revue d'Histoire Diplomatique* 122, no. 1 (2008): 61–75.

Kozyrev, Nikolai. *Zhenevskie soglashenie 1988 goda i Afganskoe uregulirovanie* [The

Geneva Accords of 1988 and the Afghan Settlement]. Moscow: Diplomatic Academy of the Ministry of Foreign Affairs of the Russian Federation, 2000.

Krakhmalov, Sergei. *Zapiski Voennogo Attashe: Iran, Egipet, Iraq, Afganistan* [Notes of a Military Attaché: Iran, Egypt, Iraq, Afghanistan]. Moscow: Russkaia Razvedka, 2000.

Kriuchkov, V. L. *Lichnoe delo: Tri dnia i vsia zhizn'* [Personal File: Three Days and a Whole Life]. Moscow: AST, Izd-vo Olimp, 2001.

Kuperman, Alan J. "The Stinger Missile and US intervention in Afghanistan." *Political Science Quarterly* 114, no. 2 (1999): 219–263.

Kuperman, Alan J., and Bearden, Milton M. "Stinging Rebukes." Online at www.foreignaffairs.com/articles/57633/alan-j-kuperman/stinging-rebukes (accessed December 27, 2010).

Kux, Dennis. *The United States and Pakistan, 1947–2000: Disenchanted Allies.* Washington, DC: Woodrow Wilson Center Press, 2001.

Kuzio, Taras. "Opposition in the USSR to the Occupation of Afghanistan." *Central Asian Survey* 6, no. 1 (1987).

Kydyralina, Zhanna Urkinbaevna. "Politicheskie Nastroenia v Kazakhstane v 1945–1985 gg" [The Political Mood in Kazakhstan in 1945–1985]. *Voprosy Istorii* 8 (2008): 64–72.

Latham, Michael. *Modernization as Ideology: American Social Science and "Nation-Building" in the Kennedy Era.* Chapel Hill: University of North Carolina Press, 2000.

——— "Redirecting the Revolution? The USA and the Failure of Nation-Building in South Vietnam." In Mark T. Berger, ed., *From Nation-Building to State-Building.* London: Routledge, 2007.

Leffler, Melvyn P. *For the Soul of Mankind: The United States, the Soviet Union, and the Cold War.* New York: Hill and Wang, 2007.

Lévesque, Jacques. *The Enigma of 1989: The USSR and the Liberation of Eastern Europe.* Berkeley: University of California Press, 1997.

——— *L'URSS en Afghanistan de l'invasion au retrait.* Montreal: Editions Complexe, 1990.

Liakhovskii, Aleksandr A.. "Inside the Soviet Invasion of Afghanistan and the Seizure of Kabul, December 1979." Trans. Gary Goldberg and Artemy M. Kalinovsky. Cold War International History Project Working Paper 51, 2007.

——— *Tragediia i doblest' Afgana* [The Tragedy and Valor of Afghanistan]. Moscow: Nord, 2004.

——— and Viacheslav Nekrasov. *Grazhdanin, Politik, Voin: Pamyat' o Akmad Shah Massoude* [Citizen, Politician, Warrior: The Memory of Ahmad Shah Massoud]. Moscow: Self-published, 2007.

Lieven, Anatol. *America Right or Wrong: An Anatomy of American Nationalism.* London: Harper Perennial, 2005.

Light, M. "The New Political Thinking and Soviet Third World Policy." *RUSI and Brassey's Defence Yearbook* 1990. London: Brassey's, 1990: 255–267.

——— "Soviet Policy in the Third World." *International Affairs* 67, no. 2 (1991): 263–280.

——— ed. *Troubled Friendships: Moscow's Third World Ventures.* London: British Academic Press, 1993.

Loth, Wilfried. *Overcoming the Cold War: A History of Détente, 1950–1991.* New York: Palgrave, 2001.

MacEachin, Douglas. *Predicting the Soviet Invasion of Afghanistan: The Intelligence Community's Record.* Washington, DC: Center for Study of Intelligence, Central Intelligence Agency, 2002. Online at www.cia.gov/library/center-for-the-study-of-intelligence/csi-publications/books-and-monographs/predicting-the-soviet-invasion-of-afghanistan-the-intelligence-communitys-record/predicting-the-soviet-invasion-of-afghanistan-the-intelligence-communitys-record.html (accessed December 27, 2010).

Magnus, Ralph H., and Naby Eden. *Afghanistan: Mullah, Marx, and Mujahid.* Boulder, CO: Westview Press, 2002.

Maiorov, Aleksandr. *Pravda ob Afganskoi voine: Svidetel'stvo glavnogo voennogo sovetnika* [The Truth about the Afghan War: Testimony of the Chief Military Adviser]. Moscow: Prava cheloveka, 1996.

Maley, William. *The Afghanistan Wars.* Basingstoke: Palgrave Macmillan, 2002.

Markey, Daniel. "From AfPak to PakAf: A Response to the New U.S. Strategy for South Asia." Council on Foreign Relations, Policy Options Paper, April 2009.

Marshall, Alex. "Managing Withdrawal: Afghanistan as the Forgotten Example in Attempting Conflict Resolution and State Reconstruction." *Small Wars and Insurgencies* 18, no. 1 (2007): 68–89.

Marwat, Fazal-Ur-Rahim. *The Evolution and Growth of Communism in Afghanistan, 1917–1979: An Appraisal.* Karachi: Royal Book, 1997.

Mastny, Vojtech. "The Soviet Non-Invasion of Poland in 1980–1981 and the End of the Cold War." Cold War International History Project Working Paper 23, 1998.

——— "The Soviet Union's Partnership with India." *Journal of Cold War Studies* 12, no. 3 (Summer 2010): 50–90.

Matlock, Jack. *Reagan and Gorbachev: How the Cold War Ended.* New York: Random House, 2004.

——— "The End of Détente and the Reformulation of American Strategy, 1980–1983." In *Turning Points in Ending the Cold War,* ed. Kiron K. Skinner. Stanford: Hoover Institution Press, 2008.

McMahon, Robert J. "The Politics, and Geopolitics, of American Troop Withdrawals from Vietnam, 1968–1972." *Diplomatic History* 34, no. 3 (2010): 471–483.

Mendelson, Sarah E. *Changing Course: Ideas, Politics, and the Soviet Withdrawal from Afghanistan.* Princeton, NJ: Princeton University Press, 1998.

——— "Internal Battles and External Wars: Politics, Learning, and the Soviet Withdrawal from Afghanistan." *World Politics* 45, no. 3 (1993): 327–360.

Menshikov, Stanislav. "Stsenarii razvitia VVP." *Voprosy Ekonomiki* 7 (July 1999): 86–99.

Merimskii, V. A. "Afganistan: uroki i vyvody." *Voenno-Istoricheskii Zhurnal* 1 (1994): 24–29.

Mitochkin, Valerii I. *Afganskie zapiski* [Notes from Afghanistan]. Saransk: Krasnyi Oktiabr, 2004.

Morgenthau, Hans J. "To Intervene or Not to Intervene?" *Foreign Affairs* (1967): 425–436.

Nablandiants, Ruben N. *Zapiski Vostokoveda* [Notes of an Orientalist]. Moscow: Luch, 2002.

Nikitenko, E. G., and N. I. Pikov. "Razvenchannyi mif o nepobedimosti plemene Dzhadran." *Voenno-Istoricheskii Zhurnal* 2 (1995): 73–79.

Oberdorfer, Don. *The Turn: How the Cold War Came to an End.* London: Jonathan Cape, 1992.

Obitchkina, Evguenia. "L'Intervention de l'Union Soviétique en Afganistan." *Revue d'histoire diplomatique* 2 (2006): 155–168.

Odom, William. *On Internal War: American and Soviet Approaches to Third World Clients and Insurgents.* Durham, NC: Duke University Press, 1992.

——— *The Collapse of the Soviet Military.* New Haven: Yale University Press, 1998.

Okorokov, Aleksandr. *Sekretnye voiny Sovetskogo Soiuza* [Secret Wars of the Soviet Union]. Moscow: Yauza, 2008.

Olesen, Asta. *Islam and Politics in Afghanistan.* Richmond: Curzon Press, 1995.

Ouimet, Matthew J. *The Rise and Fall of the Brezhnev Doctrine in Soviet Foreign Policy.* Chapel Hill: University of North Carolina Press, 2003.

Palazhchenko, Pavel. *My Six Years with Gorbachev and Shevarnadze: The Memoir of a Soviet Interpreter.* University Park: Pennsylvania State University Press, 1997.

Pankin, Boris. *The Last Hundred Days of the Soviet Union.* New York: I. B. Tauris, 1996.

Patnam, Robert G. "Reagan, Gorbachev and the Emergence of 'New Political Thinking.'" *Review of International Studies* 25 (1999): 577–601.

Pérez de Cuéllar, Javier. *Pilgrimage for Peace: A Secretary General's Memoir.* New York: St. Martin's Press, 1997.

Pikhoia, R. G. *Sovetskiy Soiuz: Istoriia vlasti, 1945–1991.* Novosibirsk: Sibirskii Khronograf, 2000.

Plastun, Vladimir, and Vladimir Adrianov. *Nadzhibulla: Afganistan v tiskakh geopolitiki* [Najibullah: Afghanistan in the Vise of Geopolitics]. Moscow: Russkii Biograficheskii Institut, 1998.

Pomper, Stephen D. "Don't Follow the Bear: The Soviet Attempt to Build Afghanistan's Military." *Military Review* 85, no. 5 (2005): 26–29.

Poullada, Leon. *Reform and Rebellion in Afghanistan, 1919–1929: King Amanullah's Failure to Modernize a Tribal Society*. Ithaca: Cornell University Press, 1973.

Prados, John. *Safe for Democracy: The Secret Wars of the CIA*. Chicago: Ivan R. Dee, 2006.

Primakov, Evgeny. *Gody v bol'shoi politike* [Years in High Politics]. Moscow: Sovershenno sekretno, 1999.

Radchenko, Sergei. *Facing the Dragons*, manuscript in progress.

Rashid, Ahmed. *Descent into Chaos: The United States and the Failure of Nation Building in Pakistan, Afghanistan and Central Asia*. New York: Viking, 2008.

——— *Jihad: The Rise of Militant Islam in Central Asia*. New Haven: Yale University Press, 2002.

——— *Taliban: Islam, Oil and the New Great Game in Central Asia*. London: I. B. Tauris, 2002.

——— *Taliban: The Story of the Afghan Warlords*. London: I. B. Tauris, 2000.

Reagan, Ronald. *An American Life*. New York: Pocket Books, 1990.

——— *The Reagan Diaries*. New York: HarperCollins, 2007.

Reuveny, Rafel, and Aseem Prakash. "The Afghanistan War and the Breakdown of the Soviet Union." *Review of International Studies* 25 (1999): 693–708.

Robinson, Paul. "Russian Lessons: We Aren't the First to Try Nation Building in Afghanistan." *American Conservative*, August 1, 2009. Online at amconmag.com/article/2009/aug/01/00030/ (accessed July 14, 2009).

Rogers, Tom. *Soviet Withdrawal from Afghanistan: Analysis and Chronology*. Westport, CT: Greenwood Press, 1992.

Ro'i, Yacov. *Islam in the Soviet Union: From the Second World War to Gorbachev*. London: C. Hurst, 2000.

——— ed. *The USSR and the Muslim World*. London: George Allen and Unwin, 1984.

Romanchenko, Iu. G. "'Pridet Vremia, Kogda Poiumut, Chto My Tam Voevali Za Svoe Otechestvo': Geopoliticheskii Aspekty Sovetski-Amerikanskogo Sopernichestva v Afganistane" ["One Day People Will Understand That We Were Fighting There for Our Fatherland": Geopolitical Aspects of the Soviet-American Rivalry in Afghanistan]. *Voenno-Istoricheskii Zhurnal* 2 (2005): 42–46.

Roy, Arundhati. *The Soviet Intervention in Afghanistan: Causes Consequences, and India's Response*. New Delhi: Associated Publishing House, 1987.

Roy, Olivier. *The Failure of Political Islam*. Trans. Carol Volk. Cambridge, MA: Harvard University Press, 1994.

——— *Islam and Resistance in Afghanistan*. Cambridge: Cambridge University Press, 1990.

Rubin, Barnett R. *The Fragmentation of Afghanistan: State Formation and Collapse in the International System*. New Haven: Yale University Press, 1995.

——— *The Search for Peace in Afghanistan: From Buffer State to Failed State.* New Haven: Yale University Press, 1995.

Rubinstein, Alvin Z. *Moscow's Third World Strategy.* Princeton: Princeton University Press, 1988.

Russia (Federation) General'nyi shtab. *The Soviet Afghan War: How a Superpower Fought and Lost.* Ed. and trans. Lester W. Grau and Michael A. Gress. Lawrence: University Press of Kansas, 2002.

Saburov, L. D. "Ispol'zovanie Material'no-Tekhnicheskoi Bazy Vospitatel'noi Raboty v Khode Boevykh Deistvii Sovetskikh Voisk v Afganistane" [Making Use of the Material-Technical Basis of the Educational Work during the Course of Military Actions of Soviet Troops in Afghanistan]. *Voenno-Istoricheskii Zhurnal* 3 (2008): 66–67.

Safronchuk, Vasilii. "Afghanistan Pri Babrake Karmale i Nadzhibulle: Dolgiy put' k Zhenevskim Soglasheniam" [Afghanistan under Babrak Karmal and Najibullah: The Long Path to the Geneva Accords]. *Aziia i Afrika Segodnia* 5, no. 1 (1997).

Saikal, Amin, and William Maley. *The Soviet Withdrawal from Afghanistan.* Cambridge: Cambridge University Press, 1989.

Salnikov, Yurii. *Kandahar: Zapiski sovetnika posolstva* [Kandahar: Notes of an Embassy Adviser]. Volgograd: Kommitet po pechati, 2005.

Sarin, Oleg, and Oleg Dvoretskii. *The Afghan Syndrome: The Soviet Union's Vietnam.* Novato, CA: Presidio, 1993.

Savranskaya, Svetlana. "Gorbachev and the Third World." In *The End of the Cold War and the Third World,* ed. Artemy M. Kalinovsky and Sergei Radchenko. London: Routledge, 2010.

——— "Voenno-Politicheskie Aspekty Okonchaniia Kholodnoi Voiny." In *Konets Kholodnoi Voiny: Novye Fakty i Aspekty,* ed. Vladislav Zubok. Saratov: Nauchnaia kniga, 2004.

Schulze, Kirsten E. "The Rise of Political Islam, 1928–2000." In Antony Best et al., eds., *International History of the Twentieth Century.* London: Routledge, 2003.

Scott, James C. *Seeing Like a State: How Certain Schemes to Improve the Human Condition Have Failed.* New Haven: Yale University Press, 1998.

Scott, James M. *Deciding to Intervene: The Reagan Doctrine and American Foreign Policy.* Durham, NC: Duke University Press, 1996.

Selverstone, Marc, ed. "Special Forum: The Politics of Troop Withdrawal." *Diplomatic History* 34, no. 3 (2010).

Shakhnazarov, G. *S Vozhdyami i bez nikh* [With Leaders and Without Them]. Moscow: Vagrius, 2001.

——— *Tsena svobodi: Reformatsia Gorbacheva glazami ego pomoshnika* [The Price of Freedom: Gorbachev's Reforms through an Aide's Eyes]. Moscow: Vagrius, 2004.

Shearman, P. "The Soviet Union and the Third Word." *Third World Quarterly* 9, no. 4 (1987): 1083–1117.

Shebarshin, L. V. *Iz zhizni nachal'nika razvedki* [From the Life of an Intelligence Chief]. Moscow: Mezhdunarodnie Otnosheniia, 1994.

——— *Ruka Moskvi* [The Hand of Moscow]. Moscow: EKSMO, 2002.

Shevardnadze, Eduard. *The Future Belongs to Freedom.* London: Sinclair-Stevenson, 1991.

——— *Kogda rukhnul zheleznii zanaves* [When the Iron Curtain Fell]. Moscow: Evropa, 2009.

Shipler, David K. "Out of Afghanistan." *Journal of International Affairs* 42, no. 2 (1989): 477–486.

Shultz, George P. *Turmoil and Triumph: My Years as Secretary of State.* New York: Scribners, 1993.

Simpson, Bradley. *Economists with Guns: Authoritarian Development and U.S.-Indonesian Relations.* Stanford: Stanford University Press, 2008.

Sinno, Abdulkader H. *Organizations at War in Afghanistan and Beyond.* Ithaca: Cornell University Press, 2008.

Skocpol, Theda. *Social Revolution in the Modern World.* Cambridge: Cambridge University Press, 1994.

Slinkin, M. F. *Narodno-Demokraticheskaia Partia Afganistana uvlasti vremena Taraki i Amina.* Simferopol: Simferopolski Gosudarstvenniy Universitet, 1999.

Sotskov, M. M. *Dolg i sovest: Zakrytie stranitsy Afganskoi voiny* [Duty and Conscience: Pages from the Afghan War]. St. Petersburg: Professional, 2007.

Strong, Robert. *Working in the World: Jimmy Carter and the Making of American Foreign Policy.* Baton Rouge: Louisiana State University Press, 2000.

Talbott, Ian. *Pakistan: A Modern History.* London: Hurst, 1998.

Taliaferro, Jeffrey. *Balancing Risks: Great Power Intervention in the Periphery.* Ithaca: Cornell University Press, 2004.

Tarzi, Shah M. "Politics of the Afghan Resistance Movement: Cleavages, Disunity, and Fragmentation." *Asian Survey* 31, no. 6 (1991): 479–480.

Topuridze, N. "*Afganistan: Voinu proigrali politiki*" [Afghanistan: Politicians Lost the War]. *Argumenty i Fakty* 39, September 30, 1989.

Ulam, Adam Bruno. *Dangerous Relations: The Soviet Union in World Politics, 1970–1982.* Oxford: Oxford University Press, 1983.

USSR-USA Summit, Washington, December 7–10, 1987: Documents and Materials. Novosti Information Agency. Moscow, 1987.

USSR-USA Summit, Moscow, May 29–June 2, 1988, Documents and Materials. Novosti Information Agency. Moscow, 1988.

Valenta, J., and F. Cibulka, eds. *Gorbachev's New Thinking and Third World Conflicts*. New Brunswick, NJ: Transaction, 1990.

Vance, Cyrus. *Hard Choices: Critical Years in America's Foreign Policy*. New York: Simon and Schuster, 1983.

Varennikov, V. I. *Nepovtorimoe, tom 5: Afganistan* [Unrepeatable, vol. 5: Afghanistan]. Moscow: Sovetskiy Pisatel, 2002.

——— *Afganistan: Podvodia itogi* [Afghanistan: In Sum]. Interview with Artem Borovik, *Ogonek*, no. 12, 1989.

——— "Sud'ba i Sovest'" [Fate and Conscience]. Interview with Boris Kurkin, Moscow: Paleia, 1993.

Vorotnikov, V. I. *A bylo eto tak: Iz dnevnika chlena Politburo TsK KPSS* [This Is How It Was: From the Diary of a Politburo Member]. Moscow: Kniga i bizness, 2003.

Waldheim, Kurt. *In the Eye of the Storm: The Memoirs of Kurt Waldheim*. London: Weidenfeld, 1985.

Wallander, Celeste A. "Soviet Policy toward the Third World in the 1990s." In *Third World Security in the Post–Cold War Era*, ed. Thomas G. Weiss and Meryl A. Kessler.

Weiss, Thomas G., and Meryl A. Kessler, eds. *Third World Security in the Post–Cold War Era*. Boulder, CO: Lynne Riener, 1991.

Weitz, Richard. "Russia Returns to Afghanistan," *Foreign Policy*, 3 November 2010. Online at afpak.foreignpolicy.com/posts/2010/11/03/russia_returns_to_afghanistan (accessed January 3, 2011).

Westad, Odd Arne. "Concerning the Situation in A.: New Russian Evidence on the Soviet Intervention in Afghanistan." *Cold War International History Project Bulletin* 8–9. Washington, DC: Woodrow Wilson Center Press, 1989.

——— *The Fall of Détente: Soviet-American Relations during the Carter Years*. Oslo: Scandinavian University Press, 1997.

——— *The Global Cold War*. Cambridge: Cambridge University Press, 2005.

——— ed. "US-Soviet Relations and Soviet Foreign Policy toward the Middle East and Africa in the 1970s." Transcript from a workshop at Lysebu, October 1–3, 1994. Oslo: Norwegian Nobel Institute, 1995.

Westermann, Edward W. "The Limits of Soviet Airpower: The Failure of Military Coercion in Afghanistan, 1979–1989." *Journal of Conflict Studies* 19, no. 2 (1999): 39–71.

Wilentz, Sean. *The Age of Reagan: A History, 1974–2008*. New York: HarperCollins, 2008.

Wolf, Matt W. "Stumbling towards War: The Soviet Decision to Invade Afghanistan." *Past Imperfect* 12 (2006): 1–19.

Woodward, Bob. *Obama's Wars*. New York: Simon and Schuster, 2010.

Young, Marilyn B. *The Vietnam Wars, 1945–1990*. New York: HarperCollins, 1991.

Yousaf, Mohammad, and Mark Adkin. *The Bear Trap: Afghanistan's Untold Story.* London: L. Cooper, 1992.

Zharov, O. "Sleptsi, Navizivavshie Sebia v Povodyri." *Aziia i Afrika Segodnia* 12 (1992): 29.

Zubok, Vladislav M. A *Failed Empire: The Soviet Union in the Cold War from Stalin to Gorbachev.* Chapel Hill: University of North Carolina Press, 2007.

——— "Gorbachev and the End of the Cold War." *Cold War History* 2, no. 2 (2002): 61–100.

——— "Why Did the Cold War End in 1989?" In *Reviewing the Cold War: Approaches, Interpretations, Theory,* ed. Odd Arne Westad. London: Frank Cass, 2000.

ACKNOWLEDGMENTS

Many people made this book possible. At the London School of Economics and at the Centre for Diplomacy and Strategy (LSE IDEAS), my academic home for most of the time I spent working on this book, my thanks go first and foremost to Arne Westad and Anita Prazmowska. A number of people at IDEAS and beyond read chapters and provided invaluable advice and support along the way; they are Dominic Lieven, David Priestland, Paul Keenan, Antonio Giustozzi, Tanya Harmer, Sergei Radchenko, Arne Hofman, Vanni Pettina, Dina Fainberg, Tom Field, Joris Versteeg, Vladimir Unkovski-Korica, Brian Becker, Rodric Braithwaite, and Taylor Asen. At George Washington University, Muriel Atkin, James Hershberg, Hope Harrison, and Svetlana Savranskaya provided commentary and advice on this project as it evolved. James Wilson graciously shared documents from the Ronald Reagan Presidential Library that became available only after my trip there. At the University of Amsterdam, where this manuscript was completed, I would like to thank my colleagues for a warm welcome that made the transition to a new institution easier and my work that much more enjoyable. Finally, I would like to thank Maurice Pinto for his enthusiastic support and encouragement.

In Moscow, Evgeny Golynkin and Alla Nikolaevna provided a welcoming home during my research trips. Historical research would be impossible without archivists, and I am particularly grateful to the staff at the Gorbachev Foundation (Fond Gorbacheva), the Russian State Archive of Contemporary History (RGANI), the State Archive of the Russian Federation (GARF), and the Russian State Archive of the Economy (RGAE). Friends and colleagues helped to locate sources and reach interviewees; I am especially grateful to Vladimir Pechatnov, Vladimir Shubin, Natalia Kapitonova, and Mikhail Lipkin. This project would have been impossible without the help of Aleksandr Liakhovskii, who has done more than anyone to record the history of the Soviet war. Sadly, he passed away in February 2009, a week before the twentieth anniversary of the Soviet withdrawal.

Conversations with a number of Afghans involved in the conflict have been very helpful to me. I would like to thank in particular the Sadeghi family, Seraya Baha, and Colonel Ahmet Muslim Hayat.

At Harvard University Press, I was fortunate to have the support of several excellent editors: Kathleen McDermott, Maria Ascher, and Sophia Saeed Khan. They have been attentive, supportive, and meticulous—everything I hoped editors would be. Thanks also to Isabelle Lewis for the maps.

At the end of the day, this project would not have been possible without the support and encouragement of those closest to me. So I must thank my grandmother, the rest of my family, and Dayna—for thinking the project was a good idea, for reminding me of this when I found it hard to believe, and for much much more.

INDEX